1/03

D0979730

Aiming for the Stars

$\frac{09}{1}$ $\frac{10}{1}$

GAYLORD MG

AIMING
FOR THE
STARS

The Dreamers
and Doers
of the
Space Age

TOM D. CROUCH

SMITHSONIAN INSTITUTION PRESS
Washington and London

Copy editor: Karin Kaufman
Production editor: Duke Johns
Designer: Janice Wheeler

Library of Congress Cataloging-in-Publication Data
Crouch, Tom D.
    Aiming for the stars : the dreamers and doers of the space age / Tom D. Crouch
      p. cm.
    Includes bibliographical references and index.
    ISBN 1-56098-386-8 (alk. paper)
    1. Astronautics—History. 2. Space flight—History.
    3. Aerospace engineers—History. I. Title.
    TL788.5.C67   1999
    629.4—dc21   99-13822

British Library Cataloguing-in-Publication Data available

A paperback reissue (ISBN 1-56098-833-9) of the original cloth edition

Manufactured in the United States of America
07 06 05 04 03 02 01 00 5 4 3 2 1

Contents

Illustrations

Tables

Preface

The book you hold in your hands offers an account of the extraordinary human beings who invented the space age, sold it to the powers that be, and managed the development of a complex technological enterprise that has shaped the history of the twentieth century—and been profoundly shaped by it. It is a short history of space flight in the form of a narrative that explores the lives and work, personalities and motivations, goals and dreams, and trials and triumphs of generations of men and women who either dreamed of pioneering the ultimate frontier or transformed the dream into reality.

A novelist could not concoct a more fascinating cast of characters. The book opens with the sixteenth-century genius who first recognized the possibility of space travel and proceeds to a handful of turn-of-the-century thinkers who pursued the dream of traveling beyond Earth with something akin to religious zeal, proved that it could be done, and inspired a generation of brilliant engineer/managers who invented the guided missile as a means of taking the first steps toward the goal of space flight.

A dream born and nursed in the hearts and minds of a small number of deeply committed enthusiasts operating on the fringes of science and society had converged with the needs of twentieth-century nation-states. Private dreams and small-scale efforts were replaced by state-sponsored programs employing millions of individuals from every corner of the nation and every level of the economy. This shift in the nature of the enterprise is reflected in the changing narrative, which opens with an emphasis on the vision of individual pioneers and closes with an account of the great projects and programs of the space age and the achievements of the astronauts and cosmonauts, the first representatives of their species to venture away from Earth, if only for short periods of time.

Aiming for the Stars tells the story of the birth and early history of the space age through the experience of those who were there. It is a tale not only of bold dreamers and triumphant human achievement but also of obsession, opportunism, Faustian bargains, and the threat of world destruction. All of which is to say that the story of our first steps on the road to an ultimate destiny out there among the stars mirrors the deep complexities of twentieth-century history.

For all of that, however, this is ultimately a tale of human beings taking their first steps away from the ancestral home. If that great getting-up morning ever does arrive, when the Sun has grown cold and we are loading the interstellar arks that will spread human seed and civilization across the galaxies, our progeny's progeny will surely remember, and celebrate, the names of Tsiolkovsky, Goddard, Oberth, and a good many of the other figures whose lives and work are described in these pages.

Acknowledgments

Aiming for the Stars has been a long time in the making. Originally commissioned in 1993 by Alexis "Dusty" Doster III and Patricia Gallagher of Smithsonian Books, it was intended to be published as a lavishly illustrated book aimed at the mail-order market. Unfortunately, that publishing program ceased operations while the book was being developed. Two years ago, Roger Launius, chief historian of the National Aeronautics and Space Administration, brought the project back to life. My thanks to Dusty, Pat, and Roger, and to Felix Lowe, retired director of the Smithsonian Institution Press. Mark Hirsch guided the process of revising and updating the old manuscript. Karin Kaufman, the copy editor, transformed a difficult manuscript into a coherent whole. I owe a special debt to Frederick C. Durant III, Frank Winter, and Michael Neufeld, colleagues new and old. Much of the strength of the book I owe to those who came before me. The shortcomings are mine alone.

Prologue

Spaceport!

Roughly three million people visit the John F. Kennedy Space Center at Cape Canaveral, Florida, each year. They come from every state in the union and from dozens of nations around the globe. Although most are first-time visitors, they come with personal memories of what occurred here: the great events of the space age have become the bench marks and touchstones of their own lives.

For some the name John Glenn conjures memories of watching history being made on a television set that had been wheeled into a school gymnasium. Their grandchildren will remember another day, thirty-seven years later, when retired U.S. senator John Glenn returned to orbit with the crew of Space Shuttle mission STS-95. Others will recall a hot July evening in 1969 when their parents kept them up late to watch the first human being set foot on the Moon.

Americans of any generation may not know the name of the current secretary of defense or the treasury, but they know who Alan Shepard was, and who took the first step on the Moon. If they are old enough, they will remember where they were when they first learned of the Challenger disaster, the way they sat on the edge of their chairs while the Space Shuttle crew of mission STS-61 struggled to repair the Hubble Space Telescope, or how thrilled they were with the photos that proved Mars Pathfinder had survived a hard landing on the surface of the Red Planet.

Over the past three decades rockets have been sent into orbit from a total of

18 spaceports maintained by the United States (4), Russia (3), China (3), Japan (2), France (1), Italy (1), the European Space Agency (1), Israel (1), and Brazil (1). The Russian facilities at Plesetsk and Tyuratam hold the world's record for the largest number of launches (1,299 and 874, respectively) from 1957 to 1990. Cape Canaveral, with 419 launches, cannot even claim to be the most active American launch facility. That honor goes to Vandenberg Air Force Base (487 launches), where the U.S. Air Force conducted most of its defense-related satellite launches during the pre-Shuttle era.

But those are only dry statistics to the visitors who pour through the gates of the Kennedy Space Center. For them, *this* is the place that comes to mind when they think of the great events of the space age. The public enters the KSC through the Spaceport Central visitor center. They will want to take a look at the exhibits in various buildings, see one of the spectacular large-format films, pay their respects at the Astronaut Memorial, stroll through the outdoor missile park, visit the gift shop and have a snack at the Orbit Restaurant. If they are like most visitors, however, the highlight of their visit will be a bus tour of the giant launch complex. They can choose from a number of excursions. The most popular includes those sites on Merritt Island associated with the Apollo and Space Shuttle programs. This is the heart of the Kennedy Space Center.

Before 1962 all launch activity occurred several miles to the east, across the Banana River, where a long, thin ribbon of sand elbows out into the Atlantic to form the most prominent cape on the Florida coast. At that time Merritt Island was still a subtropical paradise. Tangled hummocks of low brush and scrub pines, gnarled oaks bent and twisted by the constant winds, and palmetto palms still covered the northern and western sections of the island. Roughly twenty-five hundred acres (one thousand hectares) of orange groves containing 185,000 trees were clustered to the south and west. The tiny vil-. lage of Merritt was the local urban center. Two churches served the spiritual needs of the growers and small farmers, and three small cemeteries marked the last resting place of their predecessors. A handful of summer houses were scattered along the beach.

President John F. Kennedy's stirring call to send American astronauts to the Moon changed all that. The older launch facility, strung out along the barrier island that marked the Atlantic shore, would not be adequate to the task. The giant Saturn 5 rockets that would boost American astronauts to the Moon would be propelled by 5.9 million pounds (2,655,000 kilograms) of liquid

The first rocket to roar aloft from Cape Canaveral, Florida, launched on 24 July 1950, was a captured German V-2 topped with a WAC Corporal second stage. A dozen years later the decision to send Americans to the Moon paved the way for the construction of the Kennedy Space Center on nearby Merritt Island. (Courtesy of the U.S. Air Force)

hydrogen, liquid oxygen, and kerosene—roughly equivalent to 1 million pounds (450,000 kilograms) of TNT. Such a vehicle would require a launch safety zone of 3.5 miles (5.6 kilometers); enormous buildings constructed to house the rockets during assembly and testing; and new offices, machine shops, and other work spaces to support the operation. The space program had outgrown its original home.

NASA considered building its new launch complex somewhere else. A joint team created to study potential launch sites that would meet the requirements of the Apollo program considered locations in Hawaii, Texas, California, Georgia, and the Caribbean. Given the presence of the engineers, contractors, and the transportation system currently servicing launch operations at the cape, however, there was really only one decision to be made. The existing complex would continue to be used for unmanned launches, and NASA would construct an entirely new space center devoted to the Apollo Moon program across the Banana River on Merritt Island.

By the end of 1962 NASA had purchased 83,894 acres (33,558 hectares) of island land and obtained another 55,805 acres (22,322 hectares), most of it semisubmerged, from the state of Florida. The total cost of land acquisition was $71,872,000. The original site had evolved into a long line of pads sprouting from south to north along the beach. Each launch complex was designed to support a particular series of rocket or a single program. Launch vehicles were prepared in buildings set back from the beach, trucked to the launch site horizontally, and erected on the pad. Handling the giant Saturn 5 would require a very different arrangement.

The Kennedy Space Center (officially named on 29 November 1963, one week after the assassination of President John F. Kennedy) would have only two pads, 39A and 39B, united in a single launch complex. The facility would include an enormous "hangar," mobile platforms on which the rockets would be assembled and from which they would be launched, a transport system capable of moving 12 million pounds (5.4 million kilograms) of rocket and launch stand from the hangar to a pad 3.5 miles (5.6 kilometers) away, a gantry forty-five stories tall that would surround the rocket on the pad, and a communications center from which all launch activities would be monitored and controlled. As *Architectural Digest* noted, the design of Launch Complex 39 represented "one of the most awesome construction projects ever undertaken by Earthbound men."

The largest structure at the complex is known as the Vehicle Assembly Building (VAB). You can see the VAB from more than ten miles away. It rises out of the flat landscape as you pass through Gate 2C and drive north up the Kennedy Parkway. With no other buildings in the immediate neighborhood, it takes some time to realize just how large this structure is. It stands 526 feet (160 meters) tall, is 518 feet (157 meters) wide, and is 716 feet (217 meters) long. All told, the VAB covers 8 acres (3.2 hectares). The building was designed to house as many as three of the 363-foot (110-meter) Saturn 5 rockets and a Saturn 1B, all standing upright at one time. You could wheel the Statue of Liberty through the gigantic doors leading to the pad.

Every rocket launched from the complex has moved from the VAB to the pad by means of an enormous "crawler" mounted on caterpillar treads of the sort used on strip-mining shovels. Powered by twin 2,750-horsepower diesel engines driving four 1,000-kilowatt generators, each vehicle weighs 6 million pounds (2.7 million kilograms) and can move 12 million pounds (5.4 million kilograms) of rocket, held upright in a mobile launch tower, along a 3.5-mile (5.6-kilometer) crushed-stone "crawler way" at a top speed of 1 mile (1.6 kilometers) per hour. Two more diesels drive the generators that operate the steering mechanism and level the platform. In a crisis, such as the approach of a hurricane, it is possible to move a rocket from the pad back into the safety of the VAB in less than twelve hours.

Launch pads 39A and 39B stand on land that was reclaimed from the sea. Some 4.5 million cubic yards (3.42 million cubic meters) of fill were dredged from the bottom of the Banana River to create the two launch areas. Each pad is an octagon covering roughly .5 square miles (1.3 square kilometers) of land and includes a concrete hard stand complete with six mounting mechanisms to secure the mobile launcher, flame deflectors, and four underground floors housing essential systems and communications gear that links the pad to the launch-control center 3.5 miles (5.6 kilometers) to the west.

A single Mobile Service Structure (MSS) serves both pads. With its top rising above the nose of a Saturn 5, it has a base half the size of a football field; a platform 113 feet (10 meters) square at the top; and intermediate platforms from which workers load propellant, service the rocket, and conduct the final systems checks. When not in use the structure is parked some 7,000 feet (2,123 meters) from pad A.

Propellant that will be transferred to a launch vehicle is stored in huge tanks scattered around the perimeter of each pad. Vacuum-jacketed pipes, clad in stainless steel, move the propellant from the tanks to the handling system on the MSS, which feeds liquids into the tanks of the rocket. There is a burn pond to handle hydrogen boil-off, and there are holding ponds to contain other fuel and oxidizer spills.

Each pad is furnished with a 700-ton (637-metric-ton) steel flame deflector that directs the rocket exhaust along a brick-lined trench measuring 42 feet (13 meters) deep and 54 feet (18 meters) wide. During the thirty seconds after ignition, 25,000 gallons (94,750 liters) of water roar through the deflector and flame trench, reducing the level of noise and vibration and cooling critical areas of the pad.

Every Apollo mission to the Moon began at Launch Complex 39. It is, in fact, still the spot from which every American astronaut begins his or her journey into space, as the VAB has been reconfigured to prepare Space Shuttles for launch. The winged orbiter, external propellant tank, and solid rocket boosters are assembled vertically in this structure, just as the giant Saturn 5s were three decades ago. The Shuttle, which seems enormous when seen out in the open, is dwarfed by its hangar. The crawler still creeps down the path to the pad, and water still rushes through the flame deflector at ignition.

Administrative buildings, an industrial area, contractor facilities, laboratories, an airstrip, and specialized structures to house telemetry equipment, payload processing, and other essential operations have been constructed on Merritt Island over the years to support the activities taking place out on Launch Complex 39. During the period of peak Apollo activity at the cape, 26,500 men and women came to work here every day. The once-quiet communities of Titusville and Cocoa Beach were transformed into neon-bedecked boomtowns, complete with a wide selection of bars and motels, where the desk clerks and bartenders still spin Homeric tales of the glory days of Mercury, Gemini, and Apollo.

In view of the unprecedented growth that has occurred here since 1962 it is astonishing that so much has remained essentially unchanged. Oranges are still the principal agricultural product of Merritt Island. NASA purchased the groves in order to obtain control of all the land that might one day be required for expansion, then immediately leased the citrus farms back to their former owners for ten years. Moreover, through an interagency agreement between NASA and the Department of the Interior, all of the Kennedy Space Center became part of an even larger federal entity—the Merritt Island National Wildlife Refuge. Space flight is the center's primary business, but it is pursued with the interest of the native flora and fauna very much in mind.

Comprising some 140,000 acres (56,000 hectares) of land, the refuge is still home to virtually all of the animal species that were present in the nineteenth century—from bobcats and wild pigs to the ever-present raccoons, armadillos, and snakes. Alligators are especially popular with younger tourists. For many years, a pond in front of the NASA headquarters building was kept stocked with the animals. The annual Christmas bird count, undertaken by members of the Indian River chapter of the National Audubon Society, averages close to two hundred species each year. The insects continue to thrive and are in no need of protection.

Some nostalgic visitors to Spaceport Central forego the excitement of a modern space center in favor of a longer tour involving a bus ride over NASA Causeway East to the original launch area, now known as Cape Canaveral Air

**The Space Shuttle Columbia lifts off from Launch Complex 39 on 27 June 1982.
The local denizens of the surrounding Merritt Island National Wildlife Refuge have
the best view of the launch. (Courtesy of NASA)**

Force Station. After a stop at the U.S. Air Force Space Museum (the block-
house control center for the launch of both the first American satellite and the
first astronaut), the bus proceeds north up ICBM Road. The driver will suggest that his passengers keep their eyes on the east side of the road. Every
quarter of a mile or so, there is an explanatory marker, a large bronze plaque,
or a piece of sculpture commemorating one of the launch sites in use during
the Mercury, Gemini, and early Apollo programs.

Next to each marker, an old road leads several hundred yards back to the
actual pad. Visitors are not allowed back there. It is not safe, because most of
the old launch sites have been abandoned for a quarter of a century or more.
The gantries were also removed years ago. The concrete is broken, the iron is
rusted, and everything is overgrown with weeds and vines. The subtropical
climate and the salt air have transformed these places from which the first
Americans departed for space into ruins in the making.

The Atlantic beach, hidden by the thick vegetation, lies no more than a good stone throw behind the old line of pads. Walk south along that beach for a mile or so. When you reach the perimeter of Launch Complex 46, where the U.S. Navy tests its Trident 2 missiles, you will be standing on the broad, sandy point where American history was made more than a century before the Pilgrims came ashore at Plymouth Rock.

No one knows what the Ay people called this place. They were in residence here in April 1513, when Juan Ponce de León passed this way with the caravels *Santiago* and *Santa Maria de la Consolación* and the brigatina *San Cristobal*. He first came ashore fifty miles to the north and named the land La Pascua Florida (floral Passover) in honor of the Easter season and the fragrant flowers growing there. He rounded this cape one week later, having twice been forced back by a strong ocean current flowing north. Among his other achievements, Ponce de León had discovered the Gulf stream. In its honor he called this place Cabo de Corrientes, or the "cape of currents."

The name was well chosen, but it did not stick. On his map of 1564 the artist Jacques le Moyne, a member of an ill-fated French colony established farther up the coast, identified the site as Cape Canaveral, the "cape of reeds." So it would remain, the southernmost of six great capes angling out from the Atlantic Coast of the United States: Cod, Hatteras, Fear, Lookout, Romaine, and Canaveral.

Cape Canaveral breaks the long, straight line of the Florida coast, sweeping out into the ocean at the northern end of the Bahamas passage. The quickest route up the coast for ships traveling north with the Gulf stream on the first leg of a journey across the Atlantic, the passage can be a delight in good weather. In bad weather it is a narrow trap filled with reefs and bars. For a ship caught by strong winds in the passage, there is no escape to the east or west, and no good harbor in which to seek refuge.

Pedro Menéndez de Avilés, governor general of La Florida, founder of St. Augustine, and scourge of British and French interlopers, was cast ashore here in July 1572, the victim of a shipwreck. Having convinced the hostile Ay people that they were dealing with the scout for a large Spanish force following close behind, he set out on foot for St. Augustine, where, two months later, he was welcomed back from the dead.

History was to come full circle at Cape Canaveral in the most astonishing fashion. Four hundred and two years to the month after Menéndez de Avilés, the founder of the first permanent European settlement on the North American continent, was washed ashore here, three human beings lifted off from the cape on our first journey to another world. At the time of that first unscheduled European visit to the Cape of Reeds, the man who would discover the principles of celestial mechanics on which a trip to the Moon would be based was already seven months old.

1 • A Plurality of Worlds

There will certainly be no lack of human pioneers when we have mastered the art of flight. Who would have thought that navigation across the vast ocean is less dangerous and quieter than in the narrow, threatening gulfs of the Adriatic, or Baltic, or the British straits? Let us create vessels and sails adjusted to the heavenly ether, and there will plenty of people unafraid of the empty wastes. In the meantime, we shall prepare, for the brave sky-travelers, maps of the celestial bodies—I shall do it for the moon, you Galileo, for Jupiter. *Johannes Kepler (1571–1630)*

The man who can fairly be regarded as the first significant figure of the space age was only a babe in arms when Pedro Menéndez de Avilés encountered the Ay people on the beach at Cape Canaveral. Born at Weil der Stadt, Württemburg, on 27 December 1571, Johannes Kepler was one of the most remarkable citizens of an extraordinary age. Slowly and painfully a new Europe was emerging from an era of chaos, confusion, and change. During the century before Kepler's birth, a Renaissance of art and literature had swept across Italy and France, the Protestant Reformation had shattered the unity of medieval Christendom, and mariners had planted their respective flags on the shores of two continents whose very existence they had not suspected.

It was a time of profound contradictions. Bigotry, intolerance, and persecution flourished while geographic and intellectual horizons were undergoing an enormous expansion. As Europeans struggled to deal with the consequences of religious wars, unprecedented social and economic upheaval, and shifting international rivalries, they also were growing accustomed to the excitement of change and the thrill of new discovery.

Johannes Kepler, a major contributor to one of the most significant intellectual revolutions in history, involving nothing less than a rearrangement of the cosmos, personified the ironies and contradictions of his age. The philosopher Immanuel Kant, surely a creditable judge of such matters, regarded him as "the most acute thinker ever born." Yet for all of that, Kepler's life was deeply rooted in a world of darkness, superstition, and violence.[1]

His father, an alcoholic mercenary soldier with a criminal record and a penchant for wife beating, deserted the family when Kepler was seventeen. His great aunt was burned as a witch. His mother was accused of the same crime, apparently with some justification, and barely escaped the stake. Giordano Bruno, a contemporary of Kepler's, was less fortunate. He was burned alive as a heretic at the Campo di Fiori in Rome on 17 February 1600 for insisting that the stars in the night sky were suns warming "a plurality of worlds," some of which, like Earth, might shelter life.

Kepler, an aloof and unpopular neurotic who admitted to a "dog-like loathing of baths," was not a particularly pleasant fellow. The man who laid the foundation for the science of celestial mechanics was best known in his own time as a mystic, an astrologer, and a numerologist. Although he acknowledged that astrology was but "the foolish step-daughter" of astronomy, he cast horoscopes for Duke Albrecht von Wallenstein, the preeminent military commander of the time, and cheerfully predicted everything from the weather to the outcome of battles based on his observations of the heavens.

Yet if Kepler lived his life with one foot firmly planted in the past, his eyes were most certainly turned toward the future, and his mind ranged much further afield than any of the great explorers of his day. Christopher Columbus, and those who followed him, crossed oceans and began the process of exploiting the wealth of new continents while Kepler dreamed of journeys into the cosmos and the exploration of entirely new worlds. Kepler, in fact, was among the first to honestly believe that men and women might one day voyage into space.

You can scarcely dream of traveling to a place until you know it exists. For the common folk of Kepler's age, as for their ancestors, the possibility of a

Johannes Kepler (1571–1630). (Courtesy of the National Air and Space Museum, Smithsonian Institution, Photo No. 91-15201)

journey beyond Earth was unthinkable. There *was* no place beyond Earth. The stars were pinpricks in a black velvet dome, revealing the light of some mysterious fire that was beyond knowing. The heavens hung very low overhead, and the notion that someone looking up into the night sky might be gazing across a billion-mile gulf or responding to light generated some ten thousand years earlier by a distant Sun was simply inconceivable.

The more educated members of society were aware of a theory, originally proposed by Parmenides, a Greek scholar of the fifth century B.C., which described the universe as a nest of spherical shells with Earth at the center, the Sun, the Moon, and planets on intermediary spheres, and the stars on the outer shell, or *primum mobile*. Over the centuries, this simple model grew increasingly complex as additional shells were added to force the spheres carrying the Moon and planets to move in such a way as to mirror the movements actually observed in the sky.

There were dissenting visions of the cosmos. The most interesting came from the mathematician and astronomer Aristarchus of Samos (c. 310–230 B.C.). Impressed by the work of Eratosthenes of Cyrene (c. 276–194 B.C.), who developed an ingenious geometric procedure for estimating Earth's size, Aristarchus set out to measure the Sun and the Moon and to determine their distance from Earth. Although he underestimated the scale of things by an order of magnitude, Aristarchus got the proportions right, and he found it difficult to believe that a body as large as the Sun would orbit an object as small as Earth. That being the case, he proposed that the movements of the heavenly bodies could best be understood by assuming that the Moon orbited Earth, which, like the other planets, was turning on its axis while circling the Sun. It was one of those rare leaps of the imagination that define genius.

Roman historian Plutarch reported that Aristarchus was threatened with banishment for his radical notions. Perhaps so. Such is often the fate of brilliant men, and of great ideas come seventeen hundred years before their time. In any event, the notion that the earth might not be at the center of all things was too radical for most thinkers. Blessed by the wisdom of the great Aristotle, and placed in something close to its final complex form by the Graeco-Egyptian philosopher Claudius Ptolemy (second century A.D.), the geocentric, spherical universe would continue to dominate cosmological thought in both the Christian and Islamic worlds until the mid-sixteenth century. Then, suddenly, it would be gone.

Why, after a millennium and a half, did the geocentric universe give way in a single generation? A growing dissatisfaction with the accuracy of the predictions offered by the model was certainly a factor. Creaking with age and the weight of a thousand alterations, the ever-turning spheres of the old system remained slightly out of tune with the observational data. Poor translations and incorrect transcriptions could no longer excuse a failure to keep accurate track of something as simple as the precise length of a year. The thing simply did not ring true.

There was an even deeper problem, however. Few could bring themselves to believe that Ptolemy's whirling, whizzing, off-centered contraption of a universe bore the slightest resemblance to the real world. By the sixteenth century a handful of philosophers were prepared to risk charges of heresy in order to demonstrate that God had structured his cosmos in a far simpler and more elegant fashion.

Although Mikolaj Kopernik (1473–1543) did not singlehandedly destroy the Ptolemaic system, he struck the first crucial blow. Known to posterity by his Latinized name, Copernicus, he proposed nothing less than a fundamental reordering of the cosmos along the lines suggested by Aristarchus of Samos.

In the pages of *De revolutionibus orbium coelestium* (On the revolutions of the celestial spheres), one of the most influential books ever published, this Polish churchman, physician, lawyer, and astronomer removed the earth, and its inhabitants, from the center of the universe. Our planet, he suggested, was but one of six worlds orbiting the Sun. Copernicus was fully aware of the way in which the leaders of his church would react to such a radical notion. As a result, he circulated a short, preliminary presentation of his ideas among scientific colleagues as early as 1512, but refused to publish *De Revolutionibus,* a full and complete treatment of his system, until shortly before his death in May 1543. In all probability Copernicus himself never saw a printed copy of his masterpiece.

The Copernican revolution did not take Europe by storm. Galileo Galilei feared that he himself would suffer "the fate of our master Copernicus," who "in the eyes of a countless host . . . appeared as one to be laughed at and hissed off the stage." Copernicus was rebuked by religious authorities from the pope to John Calvin and Martin Luther, who reminded his followers that Joshua commanded the Sun, not Earth, to stand still.[2]

Nor did astronomers rush to accept the new model. Many, like Tycho Brahe (1546–1601), the greatest observer and celestial record keeper of the age, rejected the notion of a Sun-centered cosmos for the best of all possible reasons: the Copernican model, as originally proposed, was no more accurate or elegant than the Ptolemaic one. In an effort to increase the precision of his scheme, Copernicus had resorted to the assorted astronomical tricks pioneered by his classical predecessors, who espoused a spherical cosmos.

The work of two very different men, Johannes Kepler and Galileo Galilei, was required to complete the Copernican revolution. Galileo (1564–1642), the urbane son of a wealthy patrician family, built his early reputation on

Nicholas Copernicus (1473–1543). (Courtesy of the Library of Congress)

studies of motion and gravitation, but he would earn immortality with the telescope. He did not invent the device; that honor probably goes to the Dutch instrument maker Hans Lippershey. Nor was Galileo the first to turn a telescope toward the heavens. Thomas Harriot, the brilliant English mathematician who taught navigation to Sir Walter Raleigh, had already prepared a map of the Moon on the basis of his telescopic observations.

What Galileo did, however, was to recognize the extent to which his observations supported the Copernican theory and to publicize that fact. Peering through his "optik tube," he saw the planets transformed into disks while the stars remained points of light. Venus passed through phases. Four moons could be seen circling Jupiter. Our own Moon was not a silver disk but a real place, complete with mountains, valleys, and what looked like seas. The gossamer band of the Milky Way was resolved into countless stars. There were more stars than could be seen with the naked eye and a sense of incredible depth to the sky. In the pages of a small book, *Sidereus Nuncias* (Starry messenger), published in March 1610, Galileo announced that be had seen the wonders of the Copernican universe with his own eyes.

Far to the north, Johannes Kepler read, and he wrote to Galileo, asking how he might obtain a telescope of his own. Galileo did not respond. The two men would carry on a sporadic correspondence over the years, but Galileo was never particularly impressed with Kepler's work. Here was a man, after all, so enamored of astrology that he believed the Moon exerted an influence over the tides of the ocean. "It has always hurt me to think that Galileo did not acknowledge the work of Kepler," Albert Einstein once remarked. "That, alas is vanity. You find it in so many scientists."[3]

Galileo, who described the beauty and wonder of space, argued the Copernican case on the basis of experiment and observation. It was Kepler, the theoretical genius, who explained the first principles of the cosmos. Using the finest set of astronomical observations available anywhere, he proposed what have become known as the laws of planetary motion, which explain the motions observed in the heavens in precise mathematical terms.

It had been possible to argue against the Copernican theory, but Kepler's mathematical formulation was unassailable. After twenty-two centuries of speculation the structure of the solar system was finally revealed.

Never before or since has there been such a sudden and stunning shift in the basic intellectual framework within which human beings organize their understanding of the universe. Far from standing stationary at the center of the universe, Earth, and all of its inhabitants, was rushing around the Sun at an incredible speed, spinning on its axis like a top. Even more stunning was the sudden expansion of the universe. The most generous of the Ptolemaic astronomers had assumed that the entire cosmos was smaller than the actual orbit of Earth about the Sun. Even Copernicus had envisioned a finite universe bounded by the final, outer sphere of the stars. Now that was gone, shattered by the vision of the endless starry depths of space as seen through the telescope.

Kepler had no doubt that someday man would voyage into that great unknown. "In the meantime, we shall prepare, for the brave sky-travelers, maps of the celestial bodies," he wrote. " I shall do it for the moon, you Galileo, for Jupiter."[4] He must have relished the thought of such a cosmic voyage.

Among his many accomplishments, he can fairly be regarded as the inventor of science fiction. In all the centuries before his time there had been only one other author who had told tales of a journey to the Moon and planets. Lucian of Samosata (c. A.D. 120–180) a Syrian-Greek author, had written two fictional accounts of voyages into the cosmos. But Lucian's fantastic yarns were nothing more than traditional traveler's tales—social and political satires with an unconventional setting. They might as easily have been set in some unknown corner of the earth as in space. Kepler's approach was very different. He used fiction as a means of presenting the best science of the day in the most appealing fashion and of exploring the social implications of new ideas.

In *Somnium* (Dream), Kepler tells the story of a man who falls asleep while reading an arcane book, only to dream that he is reading yet another book, the tale of Duracotus and his mother, the "wise woman" Floxhilda. Returning home after a period of employment as Tycho Brahe's astronomical assistant, Duracotus is stunned to discover that his mother knows more about the Moon than he does. She has visited the place, carried there by demons. Duracotus persuades his mother to send him to the Moon. The demons give him a "dozing draught," an anaesthetic to protect him from the hazards of the journey, and apply a moist sponge to his nostrils so that he can breath as they ascend through the thin air. Approaching the midpoint of the journey, "conveyance becomes easier" for the demons, suggesting a loss of weight.

Our hero finds the Moon exactly as Galileo had described it, complete with incredibly tall mountains, deep canyons, craters, and plains. The creatures inhabiting the various regions of the Moon have been shaped by their environments. Like pre-Copernican Earthlings, they are convinced that their own world is fixed at the center of the universe with all other planets and moons revolving around it. Suddenly the reader awakes to find that the story within a story is only, after all, a dream.

Somnium seems to have begun as a university thesis on the Copernican system completed as early as 1593. Over the years Kepler transformed the original paper by adding transparently autobiographical elements alluding to his own relationship with Brahe and to the women in his family, who had a reputation for supernatural wisdom. In addition, he altered details to keep the manuscript scientifically up-to-date. He may have begun circulating an early version of the fictionalized manuscript among friends as early as 1609. In 1615 a copy apparently fell into the wrong hands, leading to the arrest of Kepler's mother for witchcraft. Small wonder that the author refused to allow it to be published in his lifetime. The first printed edition appeared in 1634, four years after his death.

Somnium reflects the ironies and contradictions of Kepler's life. Demons carry the hero to the Moon, yet the book is filled with not only the very best contemporary science but also some startlingly good guesses. Readers must have found the book to be an enjoyable and convincing introduction to the Copernican universe. Yet *Somnium* represents a good deal more than a simple attempt to popularize and communicate good science. Kepler wanted to excite his readers by helping them to imagine what a journey to the Moon might actually be like. Unlike Lucian, who might as well have sent his voyagers to some distant land across the seas, Kepler based his inspired speculations on what little he knew, or thought he knew. He was the first author to describe the hazards that might be encountered space—airlessness and weightlessness. Rather than telling his readers what the Moon looked like through a telescope, he stood them on its surface, gave them a guided tour of the lunar topography, and showed them Earth suspended in the sky.

Kepler did not invent the notion of life on a plurality of worlds. Lucian had his hippogypii, whereas the followers of the Pythagorean tradition, and even so good a Christian as Bishop Nicholas of Cusa, had argued that God would not have limited his creation to a single race of intelligent beings. But those individuals spoke in terms of what we would now define as parallel universes or different dimensions, not the real world of the stars and planets. Kepler broke the anthropocentric mold, carefully crafting his lunar inhabitants to fit the peculiar niches of their unique environment.

Johannes Kepler had revealed the structure of the solar system and explained the most basic rules governing its operation. Not content with that, he had thrown open the doors of the imagination, suggesting that the Moon and planets were real places, new worlds separated from Earth by the vast chasm of space. Of course few people shared Kepler's confidence that human beings would ever be able to explore those worlds. The notion that one might actually travel into the "fearful void" of space was impious and outrageous—unthinkable. And yet . . .

2 • The Call of the Cosmos

n January 1898 the *Boston Post* began running a new serial entitled *Fighters from Mars, or, The War of the Worlds in and near Boston.* Appearing daily over several months, the series was based on a new novel by the thirty-two-year-old English writer Herbert George Wells. *The War of the Worlds* had not yet appeared in a hardcover edition, but on the basis of its 1897 serialization in both *Pearson's Magazine* and the *Cosmopolitan* it was already an enormous success. The editors of the *Post* capitalized on the excitement by commissioning a new version of the story set in the Boston area. The series proved so popular that the *Post* followed it up with a sequel, *Edison's Conquest of Mars,* by amateur astronomer Garrett P. Serviss, in which vengeful Earthlings journey to Mars to even the score for the Martian invasion of our planet.

All this talk of invaders from Mars was heady stuff for sixteen-year-old Robert Hutchings Goddard. "Wells's wonderfully true psychology made the thing very vivid," he would remember, "and possible ways and means of accomplishing the physical marvels set forth kept me busy thinking." He was still thinking when, on the afternoon of 19 October 1899, his father asked him to trim a large cherry tree in the back yard of Maple Hill, their home in Worcester, Massachusetts. Armed with a hatchet, a saw, and a homemade ladder, he climbed to the top of the tree and began to cut away the dead branches.

"It was one of the quiet, colorful afternoons of sheer beauty which we have in October in New England," he noted almost thirty years later in an autobiographical fragment. "As I looked toward the fields at the east, I imagined how wonderful it would be to make some device which had even the possibility of ascending to Mars, and how it would look on a small scale, if sent up from the meadow at my feet." It was as though he could see the machine before him, complete with a weight whirling around a horizontal shaft, moving more rapidly above than below, ready to be drawn up into the sky by centrifugal force. "In any event," he concluded, "I was a different boy when I descended the tree from when I ascended, for existence at last seemed very purposive."[1]

For young Robert Goddard the episode was much closer to a religious vision than an adolescent daydream. As an adult, he would keep a meticulous record of his daily experiences in a series of diaries. Occasionally he would forget to make note of his wife's birthday, or of their anniversary, but he seldom missed 19 October. Year after year the day he had climbed the cherry tree was remembered as "Anniversary Day."

Goddard remained a fan of H. G. Wells for the rest of his life, reading new books as they were issued and rereading the ones that had meant so much to him as a boy. Goddard wrote to Wells in April 1932, anxious to let the author know what an impact he had had on at least one life. "In 1898 I read your *War of the Worlds,*" he began.

I was sixteen years old, and the new viewpoints of scientific applications, as well as the compelling realism of the thing, made a deep impression. The spell was complete about a year afterward, and I decided that what might conservatively be called "high-altitude research" was the most fascinating problem in existence. . . . How many more years I shall be able to work on the problem, I do not know; I hope, as long as I live. There can be no thought of finishing, for "aiming at the stars," both literally and figuratively, is a problem to occupy generations, so that no matter how much progress one makes, there is always the thrill of just beginning.[2]

Robert Goddard was far from home in the late fall of 1938, testing rockets in the high desert near Roswell, New Mexico. Friends and colleagues back in Massachusetts kept him up to date on the local news, including the fact that the old cherry tree in the back yard of Maple Hill had been destroyed in the great hurricane that swept across New England that year. "Cherry tree down," Goddard wrote on 10 November. "Have to carry on alone."[3]

The tree had symbolized for him the dream of space flight. Unlike the tree, however, the dream would not die: on the evening of 19 October 1939 Goddard recorded the annual entry in his diary: "Anniversary day—40th."[4] Less

Robert Hutchings Goddard sits with his mother, Fannie Louise Goddard, c. 1900. (Courtesy of the National Air and Space Museum, Smithsonian Institution, Photo No. 77-6027)

than two weeks later Orson Welles and the Mercury Theater of the Air broadcast a special Halloween dramatization of *The War of the Worlds*. Welles and his fellow actors recast the tale as a series of fictional news broadcasts, interspersed with dance music. Hundreds of Americans who tuned in after the beginning of the program assumed that the broadcast was real and fled into the streets. The ensuing panic made headlines the next day. If Robert Goddard was listening that evening, it is safe to assume that he was not fooled. He knew how the story ended.

A legion of fictional cosmic voyagers had taken flight during the more than two and a half centuries separating the death of Johannes Kepler from the first appearance of H. G. Wells's tale of blood-sucking Martian invaders. The seed, planted in 1634 with the publication of Kepler's *Somnium* in Frankfurt and Francis Hicks's translation of Lucian's *Vera Historia* in England, had grown into a flourishing literary genre.

The seventeenth-century heirs of the early fictional cosmic travelers Icaromenippus and Duracotus were carried to the Moon and planets by magic carriages, flights of slightly confused migrating birds, rockets, and even bottles of dew, rising into the air with the morning Sun. Eighteenth-century authors contributed some 250 additional tales to the repertoire of interplanetary fiction, adding balloons, spring-powered catapults, and electromagnetic machines to the list of potential space-propulsion systems. Following the pattern established by Lucian, most of these tales were satiric commentaries on contemporary life and society. The explosive growth of science and technology in the nineteenth century, however, led to a return to the Keplerian tradition of the adventurous cosmic voyage as a mechanism for building interest and enthusiasm for scientific enterprise.

Jules Gabriel Verne (1828–1905) was to make himself the master of that genre. Born in Nantes, France, he had practiced law, composed songs and sketches for the Théâtre Lyrique, and worked as a stockbroker before publishing his first book, *Five Weeks in a Balloon,* in 1863. Verne was a prolific author, producing more than one hundred individual titles. His books were all of a type: "voyages extraordinaires." Like so much of the classic science fiction that was to follow in the early twentieth century, they were, by the standard of the day, really boys' books, filled with adventurous heroes—there was seldom a heroine in sight—who accomplished prodigious feats such as traveling around the world in record time, journeying to the center of the earth, voyaging beneath the sea, and flying through the air and off into space. The heroes were of a type as well: lone geniuses, outsiders moving against the social and political tide, men with big dreams that others regarded as impossible.

Space flight, one of Verne's favorite themes, was featured in five of his novels. Two of them, *From the Earth to the Moon* and *Around the Moon,* issued in 1865 and 1870, respectively, are arguably the most important and influential science fiction novels ever published. Verne bound the two volumes together with the thinnest of plots. Overwhelmed by boredom in the aftermath of the Civil War, Barbicane, an artillerist and a leading member of the Baltimore Gun Club, proposes that his colleagues accept a new challenge: shoot-

Jules Gabriel Verne (1828–1905), the prolific and influential author whose novels extolling the wonders of science and technology inspired generations of youngsters. (Courtesy of the National Air and Space Museum, Smithsonian Institution, Photo No. 76-4123)

ing a cannon shell to the Moon. Nicholl, a skeptical rival, opposes the project. A newly arrived Frenchman, Michel Ardan, resolves the feud and inspires the group by suggesting that all three men make the journey to the Moon in a manned cannon projectile.

The story itself is scarcely more than an excuse to present a detailed plan for flying to the Moon. Verne begins by outlining the basic principles governing such a journey, explaining the fine points of ballistics and orbital mechanics, and suggesting how those requirements could be met with existing technology. Step by step, he walks his readers through the technical details of his scheme, from the design, construction, and loading of a cannon 900 feet (274 meters) long to the creation of a hollow shell capable of protecting three cosmic voyagers from the cold, airless environment of space.

Like *Somnium,* Verne's books are fictionalized science lessons. But there is something more here—a message that generations of adolescent boys would take to their hearts: no goal, however fantastic, is beyond the reach of dedicated human beings armed with the power of modern science and technology.

The popularity of Verne's novels inspired other authors to try their hands at didactic space-fiction tales. In *The Brick Moon* (1869–70) American writer Edward Everett Hale employed two gigantic flywheels to fling the first fictional Earth satellite/space station into orbit. In 1880 Percy Greg published *Across the Zodiac,* a two-volume novel describing a trip to Mars. In his 1897 classic *Auf Zwei Planeten* the German novelist Kurd Lasswitz provided a detailed description of a Martian spacecraft designed to hover in a stationary orbit over the North Pole.

By the last quarter of the nineteenth century, the call of the cosmos, enunciated early on by Jules Verne and echoed by the novelists who followed him, had reached the little village of Izhevskoye, in the Spassk district of Ryazan Province, 600 miles (960 kilometers) east of Moscow, where Konstantin Eduardovich Tsiolkovsky was born on 17 September 1857. "My parents were poor," Tsiolkovsky recalled. "My father was a failure—an inventor and a philosopher. My mother, as my father used to say, possessed the spark of talent."[5] A bout with scarlet fever at the age of nine cost the young Tsiolkovsky his hearing and left him without any hope for a normal childhood. He grew into a lonely and isolated adolescent, lost in a world of precious books.

"It . . . seems to me . . . that the basic drive to reach out for the sun, to shed the bonds of gravity, has been with me ever since my infancy," Tsiolkovsky commented in 1927. "Anyway, I distinctly recall that my favorite dream in very early childhood, before I ever read books, was a dim consciousness of a realm devoid of gravity, where one could move unhampered anywhere, freer than a bird in flight. What gave rise to these yearnings I cannot say[,] . . . but I dimly perceived and longed after such a place unfettered by gravitation."[6]

We can almost see him—a lonely boy, his life circumscribed by poverty and physical infirmities—dreaming of an absolute escape to perfect freedom. And there were books suggesting that his dream might come true. "I think the first seeds were sown by the imaginative tales of Jules Verne, which

An illustration from Jules Verne's *Around the Moon* shows how
the novel's intrepid space voyagers used rockets to control the
flight of their cannon-shell spacecraft. (Courtesy of the National
Air and Space Museum, Smithsonian Institution, Photo No.
A-4086-E)

stimulated my mind," Tsiolkovsky told an admirer many years later. "I was
assailed by a sense of longing."[7]

Inspired by his reading he began to educate himself in mathematics and
science. At age sixteen he set off for Moscow, armed with an ear trumpet and
very few rubles. Unable to pay university tuition, he attended free lectures on
chemistry, physics, and astronomy. Fortunately, he also made the acquain-

tance of Nikolai Fyodorov. The illegitimate son of Prince Pyotr Gagarin, Fyodorov was firmly convinced that human beings would find their ultimate destiny on other worlds. Recognizing a kindred spirit when he saw one, he provided Tsiolkovsky with clothes, food, and assistance in qualifying for a teaching post.

Not long after arriving in Moscow in 1873 it occurred to the young student that centrifugal force might be used to propel a spacecraft. The idea struck him with the force of a blow. "I was so worked up that I couldn't sleep all night," he wrote in his autobiography. "I wandered about the streets of Moscow, pondering the profound implications of my discovery."

He came to his senses several hours later, having recognized some flaws in the scheme. "My disappointment," he recalled, "was as strong as my exhilaration had been." Still, it was a moment he would never forget. "That night had a lasting impact on me," he insisted thirty years later. "I still have dreams in which I fly up to the stars in my machine, and I feel as excited as on that memorable night."[8] Twenty-five years before Nahum Goddard sent his son scurrying up a cherry tree back in Massachusetts, K. E. Tsiolkovsky celebrated an anniversary day of his own.

Tsiolkovsky returned home to Izhevsk in 1876, fully certified as a teacher of mathematics and physics. He would pursue that career for the next forty years, moving up through a series of country schools before accepting his final appointment to the Women's Diocesan School in Kaluga, some 100 miles (160 kilometers) southwest of Moscow. Tsiolkovsky took his duties seriously but refused to abandon his research. Before and after school each day he put in long hours experimenting with steam engines and wind tunnels and preparing articles on all-metal dirigibles and "aviforms," or airplanes. Those schemes paled in significance, however, before the old dream of cosmic voyaging. "The idea of traveling into outer space," he recalled, "constantly pursued me."[9]

Tsiolkovsky recorded his first serious thoughts on space flight in the pages of a small notebook. The earliest entries, dating to 1878–79, describe experiments in which chicks, mice, and insects were subjected to rides on a homemade centrifuge. By the spring of 1883 he had begun the much more difficult business of analyzing the requirements of space flight in precise mathematical terms. What speeds would have to be achieved in order to escape Earth's gravity, to orbit at various altitudes, to reach the Moon? With answers to such questions in hand, he would have a set of requirements against which to measure the performance of potential space-propulsion systems.

He concluded that there were only two practical means of achieving space flight. Jules Verne had been correct. It was theoretically possible to shoot an artillery shell into space, but the disadvantages of such an approach to human space flight were overwhelming. The cannon would have to provide all of the energy required for the journey at the moment of firing, and the acceleration imposed by such an explosion would crush the occupants of the shell before it left the barrel of the gun. The air resistance encountered by the shell as it rushed through the dense lower levels of the atmosphere at maximum velocity would generate intense heat and pressure. Finally, once it was fired, the shell would be caught in the grip of ballistic forces beyond the control of its occupants.

The only other possibility was to develop a reaction motor. Such a propulsion system would operate on the basis of Isaac Newton's third law of motion, which states that for every action there is an equal and opposite reaction. Theoretically, a reaction power plant would function in space, for the principle did not require an atmosphere against which to react or push. The simple act of projecting matter away from an object would move that object a proportional distance in the opposite direction.

Moreover, a reaction-propelled vehicle would avoid most of the problems inherent in Verne's cannon-launch system. It would rise slowly through the dense lower atmosphere, picking up speed as it gained altitude. The power plant could be stopped, restarted, and throttled in flight, and the direction of the thrust could be altered so as to control the motion of the spacecraft. The engine could even be fired in the direction of motion to slow the craft for re-entry and landing. Theoretically, a reaction engine would operate at peak efficiency in the vacuum of space, where there would be no air resistance and nothing to interfere with the expansion of exhaust gases from the nozzle.

The theoretical advantages of reaction propulsion were clear, but the technical problems of designing such an engine loomed large. Tsiolkovsky began by developing a mathematical formula for analyzing the performance of a reaction-propelled vehicle. After thinking through all of the elements that should be part of such an equation he determined that the performance of such an engine was a function of the total thrust, measured in pounds, divided by the amount of propellant consumed, measured in pounds per second. Expressed in seconds, this measurement, now called "specific impulse," remains the standard gauge of performance for rocket engines and propellants.

Having determined the speed required for space flight as well as a means of analyzing engine performance, Tsiolkovsky had a yardstick with which to measure the efficiency of various reaction-propulsion systems. By 1883 he had produced his first design for a space vehicle, a spherical craft fitted with gyroscopes and propelled by the recoil of a repeating cannon. That design, however, was quickly rejected as hopelessly inefficient, leaving the rocket as the only remaining choice.

Since appearing in China six and a half centuries earlier, black-powder rockets bad been used as weapons, fireworks, and signal launchers, as well as serving as carriers of rescue lines to ships that had run aground. Theoretically, these simple devices would function in the vacuum of space, because their black-powder propellant included both a fuel to burn and a chemical oxidizer to support combustion. The difficulty, of course, was that none of the available solid propellants produced a sufficiently high specific impulse to propel a rocket to orbital velocity. Moreover, once ignited, a solid rocket motor could not be stopped, restarted, or throttled.

It required a considerable leap of the imagination to recognize in liquid propellants a solution to the problem. Some potential propellant combinations, such as liquid-hydrocarbon fuels and oxidizers such as nitrogen peroxide, were, however, readily available in the late nineteenth century. Pedro Paulet, a Peruvian chemical engineer, later claimed that while a student in Paris from 1895 to 1897 he actually had built a liquid-propellant rocket engine producing two hundred pounds of thrust (forty-five newtons) using just those propellants. Tsiolkovsky, who was interested simply in demonstrating the possibility of space flight, selected the ideal propellants: liquid hydrogen and liquid oxygen.

After ten years of thought and research the Russian schoolmaster produced a manuscript entitled "The Exploration of Space with Reaction Propelled Devices," the original version of which appeared in a 1903 issue of a science journal with a very small circulation. In this classic treatise Tsiolkovsky provided the first theoretical evidence for the possibility of achieving space flight with a liquid-propellant rocket.

Tsiolkovsky had already completed a general-arrangement drawing of a teardrop-shaped spacecraft powered by a liquid-propellant engine, but it was not included in the 1903 publication of his paper. Appearing in subsequent

Russian space pioneer Konstantin Eduardovich Tsiolkovsky (1857–1935) and a young friend. (Courtesy of the National Air and Space Museum, Smithsonian Institution, Photo No. 81-13363)

editions of the treatise, the drawing included some features that were to prove very important to the future of rocketry, such as steering vanes set in the exhaust and the use of propellants to cool the combustion chamber, a technique that would become known as regenerative cooling.

Tsiolkovsky was to spend the rest of his career refining his plans for a basic "cosmic spaceship." He considered the problems of propellant flow and pumps, steering mechanisms, and life-support systems. He designed multistage launch vehicles and winged spacecraft capable of gliding back to a landing on Earth. He analyzed a full range of other, less exotic fuels that might be used with liquid oxygen and dreamed of propulsion systems that would harness electromagnetic and atomic energy.

A standard Soviet bibliography of Tsiolkovsky's writings lists eighty-four major published works. Scattered among the serious scientific and technical studies are a surprising number of science fiction stories. Having been inspired by Jules Verne, Tsiolkovsky never doubted the importance of speculative fiction. "First, inevitably, the idea, the fantasy, the fairy tale," he commented in 1926. "Then, scientific calculation. Ultimately, fulfillment crowns the dream."[10]

Tsiolkovsky published his first science fiction story, "On the Moon," in 1892 and continued to churn out tales of space travel for the rest of his life. He used fiction as a means of exploring the technical and social implications of new ideas. The possibility of Earth-orbiting space stations, the harnessing of solar power to generate electricity, the settling of space colonies, the means of producing artificial gravity aboard a spacecraft, the use of green plants to maintain a habitable atmosphere, and the potential of hypergolic, or self-igniting, propellants were but a few of the innovations suggested in his science fiction stories.

Largely ignored during his most creative years, Tsiolkovsky was showered with honors following the Bolshevik Revolution of 1917. A regime that regarded itself as the wave of the future could only take pride in such a man. He was inducted into the Soviet Academy of Sciences, granted a generous pension, and allowed to retire from teaching in order to concentrate on his research and writing. Appreciative of these awards, he congratulated Stalin on having "brought recognition to the works of self-taught persons." The patriarch of cosmonautics died peacefully at his home in Kaluga on 19 September 1935, two days after his seventy-eighth birthday.

Tsiolkovsky left behind a powerful legacy within the USSR. The Soviet space triumphs of the late 1950s and early 1960s were the work of two generations of men and women who had grown up revering the man who established space flight as a theoretical possibility. He was virtually unknown outside the Soviet Union until the late 1920s, however, by which time scientists and technicians in other nations had discovered for themselves the key elements of Tsiolkovsky's work.

In 1919, the year the Bolshevik government honored Tsiolkovsky with membership in the Soviet Academy, the Smithsonian Institution published a sixty-nine-page pamphlet entitled *A Method of Reaching Extreme Altitudes,* the work of a thirty-seven-year-old physicist from Worcester, Massachusetts. Although Robert Goddard had never heard of Tsiolkovsky, the two men had a great deal in common. Both were the sons of would-be inventors, and both had grown up lonely and isolated as a result of severe childhood illnesses. Voracious readers, each had cultivated a special passion for tales of interplanetary travel, and at the age of sixteen each experienced a personal vision of space flight. They would both retain a mystical faith that human destiny lay out among the stars. Neither Tsiolkovsky nor Goddard was particularly comfortable working with colleagues. They were loners who would inspire young engineers but not cooperate with them. Finally, both men were dreamers in the most literal sense. Tsiolkovsky delighted in dreams of weightlessness and freedom that he equated with the sensation of space flight and relived his first youthful vision on the streets of Moscow in his dreams.

For his part, Goddard occasionally found the answers to technical problems in his dreams. In July 1915 he was deeply engaged in experiments with solid-propellant rockets and enjoying an H. G. Wells novel, *The First Men in the Moon.* On the night of 8 August he dreamed that he himself had flown to the Moon. The details of the dream remained clear enough for him to draw a sketch of his spacecraft when he awoke the next morning. "Was cold," he noted in his diary; "took photos of earth with small Kodak while there—two for stereoscopes. . . . Not enough oxygen when I opened my helmet to see if so. . . . Light was rather dim." And he added one final detail: "Set off red fire at prearranged time so all [on Earth] could see it."[11]

Within three months, Goddard had begun a series of experiments to determine how much flash powder would have to be exploded on the Moon to be visible from Earth. He published the result of his tests (2.67 pounds of Victor flash powder) in *A Method of Reaching Extreme Altitudes.*

For the rest of his life, Goddard would spend his sleeping, as well as his waking, hours in pursuit of his primary goal. Often, as in the case of the flash powder, his dreams were vivid and extraordinarily detailed. "Dreamed last night of having drops of gasoline in oxygen gas, ignited," reads a 1938 diary entry, "and passing to the place of ignition coated with a liquid of low vapor pressure and noninflammable, so that there was almost instantaneous generation of heat, but no danger of flashback."[12]

Yet if Goddard and Tsiolkovsky had much in common, there were some interesting differences between them as well. Tsiolkovsky, a self-taught man with little formal education, earned his reputation as a theorist, whereas Goddard, armed with the best scientific training his country could offer, made his primary contributions as an experimenter.

Graduating as valedictorian of his high-school class, Goddard earned a bachelor of science degree at Worcester Polytechnic Institute in 1908, and both a master's and a Ph.D. in physics from Clark University, also in Worcester, in 1910–11. He spent the next two years as an honorary fellow at Clark and then worked as a research instructor at Princeton University, returning in 1914 to Clark, where he would remain a professor of physics for most of the rest of his career.

As a graduate student, Goddard concentrated on research that ultimately had great practical value for the new field of wireless communication. Once he was free to choose his own research topics, however, he immediately returned to his old dream of space flight. While still in high school, he had submitted an article, "The Navigation of Space," to the editor of *Popular Science News,* had designed a multistage spacecraft propelled by nested cannons, and had filled his notebooks with thoughts on "The Habitability of Other Worlds," atomic energy, radio waves, solar energy, the potential hazards of space flight, and a host of other topics. In college he bombarded English professors with science-fiction stories and themes, including "On the Possibility of Navigating Interplanetary Space."

By the summer of 1909 Goddard had set out on the path blazed by Tsiolkovsky a quarter of a century earlier. He computed the velocities necessary to escape Earth's atmosphere and to achieve Earth orbit, and developed for himself the classic formulas for analyzing the performance of a reaction engine. He considered the potential of liquid propellants (including liquid oxygen and hydrogen), proposed regenerative cooling, and developed nozzle designs that would promote the efficient expansion of exhaust gases. By 1915 he had patented an electrical-propulsion system.

An experimenter by nature, Goddard had access to the well-equipped machine shop and laboratory facilities of a great American university. Seizing the opportunity, he tested the efficiency of solid propellants, comparing the thrust produced by a particular explosive compound with the ideal heat energy generated by burning the same material in a calorimeter. Then he set to work devising exhaust nozzles that would further increase thrust. In an effort to prove the theoretical possibility of operating a reaction engine in space he

Robert Goddard, Clark University's "Moon-going" professor of physics, offers a 1924 chalk talk on the basics of space travel. (Courtesy of the National Air and Space Museum, Smithsonian Institution, Photo No. 73-1274)

conducted reaction propulsion experiments in a pipe from which the air had been evacuated.

By 1915 Goddard had demonstrated to his own satisfaction that space flight could be achieved with the solid propellants then available. Faced with the short burning times associated with existing solid propellants, however, he designed a "multiple-charge" mechanism that would feed one block of propellant after another into position for ignition. To introduce his more traditional colleagues to the notion of space flight, he offered a departmental colloquium, "How to Reach High Altitudes," and completed a paper that outlined the results of his experiments and pointed to the military and scientific potential of a solid-propellant rocket capable of carrying scientific instruments to record altitudes.

Goddard had entered physics at a time when the field was bursting with new and revolutionary ideas. Both A. A. Michelson and Ernest Rutherford, Nobel laureates in physics and chemistry, respectively, had lectured at Clark.

The members of the university's physics department must have wondered why their talented young colleague was expending his time and energy on an abstract and useless engineering problem. (In view of Goddard's penchant for dreaming that question might perhaps have been addressed to Sigmund Freud, who also visited Clark during this period.)

Undeterred by any negative reaction on the part of his fellow physicists, Goddard wrote to the Smithsonian Institution on 27 September 1916 to inquire about the possibility of funds for his rocket experiments. It marked a turning point in his life, for the letter came to rest on the desk of Charles Greeley Abbot, one of the few physicists in the nation certain to take an interest in the proposal.

Director of the Smithsonian Astrophysical Observatory, Abbot had spent most of his career struggling to study solar radiation from the bottom of an ocean of air that filtered, masked, and absorbed the Sun's energy. His efforts to overcome this handicap, which ranged from deploying instrumented balloons to setting up mountaintop laboratories, had been less than successful. Goddard's proposal offered the hope that Abbot might soon be able to loft his instruments to the very top of the atmosphere.

After discussions with Abbot, Charles D. Walcott, secretary of the Smithsonian, replied to Goddard on 5 January 1917, praising the clarity of his proposal, acknowledging the importance of his project, and making an initial grant for five thousand dollars. "I think that's the most wonderful thing I ever heard of," Goddard's mother remarked. "Think of it! You send the Government some typewritten sheets and some pictures, and they send you $1,000, and tell you they are going to send four more."

In time Mrs. Goddard's son would become a master fund raiser, obtaining a grand total of $209,940 in grants between 1915 and 1941. In fact, he was one of his generation's best-funded American scientists working outside an observatory or major laboratory. As promised, he used this first Smithsonian installment to further his work on solid-propellant rockets.

With America's entry into World War I, Goddard took a leave of absence from his teaching duties and, at the request of the Smithsonian, went to work for the U.S. Army Signal Corps and the Ordnance Department. Using the laboratory facilities both at Clark and at the Mt. Wilson Observatory in California, he developed a series of tube-fired rocket projectiles that were successfully demonstrated for army officials at the Aberdeen Proving Ground in Maryland just five days before Armistice.

His military obligations behind him, Goddard finished the manuscript that he had been working on for the past five years and submitted it to the Smithsonian, which published it in 1919 as the seventy-first volume in its Miscellaneous Collections. *A Method of Reaching Extreme Altitudes* contains nothing that compared with the livelier passages in Tsiolkovsky's work. It was an engineering study, filled with quadratic equations and tabular data, a treatise designed to prove that a solid-propellant rocket could carry instruments into the upper atmosphere.

Goddard did his best to understate the more sensational aspects of his study, confining his thoughts on the possibility of far more efficient liquid propellants to a footnote and not even mentioning the possibility that human beings might one day ride on a rocket. The final few pages of the paper were devoted to "Calculation of Minimum Mass Required to Raise One Pound to an 'Infinite' Altitude"—the idea of sending a multistage rocket to the Moon. He concluded that "these developments involve many experimental difficulties, to be sure but they depend upon nothing that is really impossible."[13]

Goddard's valiant efforts to preserve his scientific dignity proved a dismal failure. A *Boston Herald* headline announced that a "New Rocket Devised by Prof. Goddard May Hit Face of Moon," and the *Boston American* heralded the appearance of a "New Jules Verne." Indeed, the shy professor and his "Moon-going Rocket" had captured the attention of the fourth estate, from the *New York Times* to the *San Francisco Chronicle*. A Bronx promoter urged Goddard to consider the Starlight Amusement Park as a potential launch site, and a Hollywood agent cabled a request: "Would be grateful for opportunity to send message to Moon from Mary Pickford on your torpedo rocket when it starts."[14]

But there were compensations to offset the disadvantages of celebrity. There were notes of congratulations from old friends such as Karl Compton, a rising star of American physics. Goddard also learned that George Ellery Hale, one of the nation's foremost astronomers, was much impressed with his work. Even more encouraging was a handful of letters from serious spaceflight enthusiasts. The first of these, dated 31 March 1920, came from Robert Esnault-Pelterie, a leading French aviation pioneer. REP, as he was known to his friends, had become interested in space flight as early as 1907, offering his first lectures on the subject in 1912. At the time of his letter to Goddard he was collecting information for an encyclopedic volume to be called *L'Astronautique,* which would appear in 1930. He congratulated Goddard on his work, requested an autographed copy of the Smithsonian monograph, and plied the young physicist with questions. It was the beginning of a long and friendly correspondence between the two men.

Goddard was less familiar with the work of Hermann Oberth, who wrote to him on 3 May 1922 to request a copy of the Smithsonian publication. Born at Hermannstadt in the Transylvania region of Romania on 25 June 1894, Oberth was eleven years old when his mother presented him with Jules Verne's Moon novels, "which," he later said, "I had read at least five or six times and, finally, knew by heart." Young Oberth undertook a mathematical analysis of the journey Verne described and concluded, sadly, that "the travelers inside the missile would have been crushed without pity by the enormous acceleration." Casting about for alternative modes of space propulsion, he considered both a magnetic launcher and a machine propelled by centrifugal force before focusing on reaction propulsion.[15]

Oberth's father, a surgeon, encouraged him to pursue medical studies at the University of Munich. Once enrolled, however, the young man opted for courses in physics and astronomy. Enlisting in the Austro-Hungarian army during Word War I, Oberth was wounded in the fighting on the eastern front and ended the war as an ambulance attendant. In addition he found the time to develop a plan for a long-range rocket weapon using liquid-air and dilute-alcohol propellants, but the scheme was rejected by German military authorities.

After the war Oberth continued his education, attending universities in Klausenburg (Romania), Munich, Göttingen, and Heidelberg. His doctoral dissertation, a theoretical proof of the possibility of space flight, was rejected by the Heidelberg faculty in 1922. "Never mind," he rebuked them. "I will prove that I am able to become a greater scientist than some of you, even without the title of doctor."

In 1923 he accepted a high school teaching post and borrowed money from his wife's savings to pay for the publication of his thesis, which appeared under the title *Die Rakete zu den Planetenraumen* (The rocket into interplanetary space). Oberth packed a wealth of information into the ninety-two page book, providing theoretical proof of the possibility of space flight; presenting the formulas for analyzing rocket performance; describing the design of a two-stage, liquid-propellant, high-altitude research rocket; and discussing the requirements for sending human beings into space. Just as Oberth's manuscript was ready for the printer, the author chanced to read a short news

Surrounded by props at the UfA movie studio in Berlin, where he was working as a technical consultant on a space-flight film, rocket pioneer Hermann Oberth takes a break from his struggle to build a working rocket. (Courtesy of the National Air and Space Museum, Smithsonian Institution, Photo No. A-3894)

article describing the work of Robert Goddard. Stunned to discover that he did not have the field to himself, Oberth wrote to Goddard at once, requesting a copy of *A Method of Reaching Extreme Altitudes.*

Oberth must have breathed a sigh of relief when he finally read the Smithsonian publication. Goddard had focused on the design of a solid-propellant rocket, passing rather quickly over the potential of liquid propellants. He had not even mentioned the possibility of sending human beings into space. Oberth prepared a short appendix for *Die Rakete,* calling attention to Goddard's work but underscoring his own independence and the fact that he had carried the matters further than had the American.

It was Goddard's turn to be stunned when, in the summer of 1923, he received a copy of Oberth's book. He had done his best to get back to work following the wave of publicity that had washed over him in 1920. Armed with an additional Smithsonian grant and with permission to use the U.S. Navy's facilities at the Indian Head Powder Factory in Maryland, he continued his experiments with solid-propellant rocket weapons for a time before turning his full attention to liquid propellants. When his copy of *Die Rakete* arrived, Goddard's workbench was cluttered with bits and pieces of pumps and tubing and his notebooks were filled with the details of experiments aimed at the production of a liquid-propellant rocket engine.

Now, Goddard explained to Secretary Walcott, here was this fellow who "sets forth the general method of achieving space flight as his own (as he claims to have been working independently), and which supposedly demonstrates that what I have been working upon cannot be developed for the ultimate uses for which his device can be applied." He continued: "I do not wish to open the question of priority and thrash out all the phases of the matter in public, certainly not at the present time when I am endeavoring to have a model completed and tried; yet I do not care to leave unanswered the claim that I have missed the most interesting part of the whole proposition, when, as a matter of fact, Oberth has himself overlooked a few important matters necessary for realizing 'Planetenraumen.'"[16]

Goddard sent the Smithsonian secretary and the trustees of Clark University fresh copies of his latest reports, along with notes establishing his priority in the field of liquid-propellant rocket research. Rather than raising the issue directly with Oberth, he went back to work, determined to allow his results to speak for themselves.

At noon on Sunday, 6 December 1925, he opened the valves to the liquid-oxygen and gasoline tanks supplying an engine mounted on a stand in the physics laboratory at Clark. A popping, crackling roar filled the building for twenty-seven seconds. During the last ten seconds of the run, the engine was producing a thrust in excess of twelve pounds. As Goddard noted that evening, "This was the first test in which a liquid-propellant rocket operated satisfactorily and lifted its own weight." The "extreme altitudes" beckoned.[17]

3 • *Raketenrummel*

obert Hutchings Goddard did not have any classes on Wednes-
day, 16 March 1926. Early that morning, he and an assistant,
Henry Sachs, loaded two liters of liquid oxygen into the professor's Chevrolet coupe and drove to an Auburn, Massachusetts, farm owned by a distant cousin, "Aunt" Effie Ward. The first order of
business was to lug sections of pipe, tools, and a spindly rocket only ten feet
(three meters) long to an open area bordering a cabbage patch. Their launch
preparations were almost complete by the time Goddard's wife, Esther, and
Percy Roope, an assistant professor whom Goddard had trained, arrived
shortly before noon.

While the others huddled behind a simple sheet-iron barricade, Henry
Sachs stood by the pipe-frame launch stand, armed with a blowtorch
mounted on the end of a broomstick. He directed the flame onto a thin igniter
casing extending from the top of the rocket. Match heads inside the casing ignited a charge of black powder. When smoke began issuing from the device,
Sachs quickly closed a pressure-relief valve on the liquid-oxygen tank and
lowered the blowtorch to the base of the rocket, where he lit a wad of alcohol-
soaked cotton in a small pan beneath the liquid-oxygen tank.

As Sachs ran toward the shelter, Goddard opened a valve connecting a
liquid-oxygen cylinder to the propellant tanks, pressurizing the system and
forcing a quantity of liquid oxygen and alcohol up into the combustion chamber, where they were ignited by the charge of black powder. With the engine

running and the rocket tethered to the ground by the oxygen hose extending back toward the shelter, Goddard pulled a lever that simultaneously broke the connection and opened a valve through which liquid oxygen entered an outer chamber heated by the burning cotton. This vaporized oxygen served to pressurize the propellant tanks during flight.

The rocket was sputtering away on the stand when Esther Goddard began filming the action with a hand-held movie camera that operated for seven seconds on a single winding. She was rewinding it at the very moment that the rocket finally lifted off and climbed to its peak altitude of 41 feet (12.5 meters). The 2.5-second flight was over by the time she started filming again. Robert Goddard would always playfully tease his wife about her failure to record, for posterity, the first flight of the world's first liquid-propellant rocket.

"It looked almost magical as it rose," Goddard noted in his diary. "It almost seemed to be saying," he continued, "I've been here long enough; I think I'll be going somewhere else, if you don't mind." Goddard reported in his diary that Esther had remarked that the rocket looked like "a fairy or an aesthetic dancer."[1] Esther later admitted that her husband had remembered his own comment with advantage. The actual words he spoke for his rocket were, "I think I'll get the hell out of here!"[2]

Goddard continued to build, static test, and launch his rockets at the farm in Auburn for the next three years. Although his record altitude remained under two hundred feet (sixty-one meters), he gained much useful experience in the design and construction of pumps, valves, combustion chambers, nozzles, igniters, and the other bits and pieces that went to make up a rocket. The steady, comfortable pace of his research was broken on 17 July 1929, when a new rocket left the launch tower with its usual roar and reached an altitude twice as high as that achieved in 1926. As Goddard and his crew were packing up for their return to Worcester, they heard the wail of an approaching siren. Summoned by neighbors who had been frightened by the noise, a police car, an ambulance, and an escort of several automobiles driven by reporters turned into Aunt Effie's drive. Robert Goddard was back in the news. His days of flying from the cabbage patch were over.

Goddard had worked hard to keep himself out of the news. In his view, a scientist pursued an avenue of research until it was completed and then published his results. There was nothing to be gained from rushing into print, and he had no intention of providing technical assistance to other potential rivals.

As a result, no one outside his immediate circle of associates, except for the officials at Clark and the Smithsonian who had read his confidential reports, was aware that he actually had built and flown liquid-propellant rockets.

In spite of his best efforts, however, Goddard was firmly entrenched as a minor celebrity by the mid-1920s. Articles about his work appeared in newspapers and magazines in both the United States and Europe. He received a steady stream of letters, such as one from Montevideo, Uruguay, that praised his "rocket of worldwide fame," and every few months he could expect to hear from yet another volunteer for a flight to the Moon.[3]

The growing number of requests for information about his work was evidence that a modest wave of enthusiasm for space flight was gathering momentum. Indeed, the publication of his own book in 1919, of Oberth's *Rakete* in 1923, and of a new and expanded 1924 edition of Tsiolkovsky's 1903 treatise marked an important turning point in the history of space flight. Inadvertently, the three pioneers had planted the seeds of a curious social movement that would pave the way for the coming of the space age.

Tsiolkovsky, Goddard, and Oberth had aimed their work at colleagues in the scientific community, yet they found their most receptive audience among a handful of young idealists who banded together to form space-flight societies and experimental rocket clubs in Europe and the United States between 1924 and 1940. Small, poorly funded, and often short-lived, these organizations served as the entry point and training ground for a core group of brilliant young engineers who, inspired by the combination of an extraordinary technical challenge and the dream of a utopian tomorrow based on travel beyond the atmosphere, would pursue the dream of space flight with an almost religious zeal.

Although little bands of rocketeers cropped up in a number of countries between the world wars, the level of general interest in the possibility of space flight varied enormously. In the United States and Great Britain the dream of voyaging through the cosmos attracted relatively few adherents and generated only a mild curiosity on the part of the public. It was literally Buck Rogers stuff, suitable for movie thrillers, pulp magazines, and the comic strips, but not something that a sensible person could take seriously. Young rocket experimenters in those nations found little support for their work.

In the Soviet Union and Germany, however, the situation was markedly different. The number of genuine enthusiasts, although still relatively small, was larger than in other nations. Rocket and space societies sprang up and faded away with some regularity, and their activity drew more attention in the popular press. Serious books on space flight appeared in greater numbers, and the best of them sold fairly well. If his correspondence is any indication, the name of Robert Goddard was better known in Germany and the USSR than it was in the United States. In these two very different nations—Germany and the Soviet Union—politics, economics, and culture had paved the way for the coming of the rocket.

A sustained, serious, and widespread interest in space flight developed first in the Soviet Union. The possibility of space travel seemed to hold a genuine appeal for first-generation Bolshevik technocrats, who defined themselves as harbingers of the future and envisioned a utopian social order based on a foundation of advanced science and technology dedicated to achieving the goals of an enlightened socialist state. One year after taking power the Bolsheviks republished Alexander Bogdanov's 1908 science fiction novel *Red Star*, in which the hero, a Russian Marxist, discovers a socialist paradise on Mars. The book would go through at least six additional printings. In 1919 Lenin and his followers rescued the aging Tsiolkovsky from obscurity and installed him as a national hero. The very fact that a Russian could claim priority as a space-flight theorist was another key element underlying tacit government support of activity in this field.

In 1920 Fridrikh Arturovich Tsander, one of Tsiolkovsky's most ardent disciples, was startled to see Lenin in the audience as he addressed a Moscow meeting of the Provincial Council of Inventors on the subject of space flight. At the end, Tsander remembered with pride, "Lenin shook my hand strongly, wished me success in my work, and promised support." Tsander, a native of Latvia, would keep faith with a regime that boasted such a farsighted leader. When questioned by a reporter as to why he wanted to visit Mars, the space pioneer pointed out that the ancients had regarded the planet as a red star, and this, he added proudly, "is the emblem of our great Soviet army."[4]

Over the next two decades, official Soviet publishers brought out a number of major works on space flight, including Yuri Kondratyuk's *Conquest of Interplanetary Space* (1929), Tsander's *Problems of Reactive Flight* (1932), Valentine Glushko's *Rockets: Their Construction and Utilization* (1935), the proceedings of a variety of symposia on rocketry and jet propulsion, and translations of important foreign works on the subject. One volume, Jakob Isidorovitch Perelman's *Interplanetary Travels*, went through ten editions and sold 150,000 copies.

N. A. Rynin's *Interplanetary Communications* was one of the most impressive and significant products of early Soviet state publishing, and illustrates the regime's commitment to the support of space-travel studies. A nine-volume encyclopedia issued between 1928 and 1932, Rynin's masterpiece offered detailed information on everything from myths, legends, and fictional accounts of space travel to the latest information on rocket theory.

The masters of the Soviet cinema also were quick to jump aboard the space-flight bandwagon. In 1924 director Yakov Protazanov filmed the classic *Aelita,* in which a group of Soviet cosmonauts journey to Mars and mount a revolution to overthrow a repressive regime. With expressionistic sets designed by Alexander Exter of the Tairov Theatre, *Aelita* offered a perfect example of the way in which the excitement of space travel could be harnessed to present a useful political message. Other space travel films followed, including a 1935 Mosfilm version of Tsiolkovsky's short story "Cosmic Journey."

By 1924 the USSR had officially sanctioned rocketry and space flight. In that year the Soviet government established the Central Bureau for the Study of the Problem of Rockets (TsBIRP), and the young officers studying at the N. E. Zhukovsky Air Force Academy organized the Interplanetary Flight Study Society. With government approval, Tsander took the lead in expanding this air force group into a broader civilian organization known as the All-Union Association for the Study of Interplanetary Communication (OIMS), headquartered in Moscow. Although short-lived, Tsander's organization at one point attracted some 150 to 200 members.

The following year Dimitri Alexander Grave established the Society for Space Study in Kiev. Like members of the OIMS, the Kiev enthusiasts arranged lectures, debates, and classes on space-flight topics. In addition the group sponsored the world's first public exhibition on the subject. Another group, the Interplanetary Section of the Association of Inventors, gathered information, models, and photographs from around the world for a much larger and more impressive exhibition, which was held in Moscow in the spring of 1927.

By 1930 official and unofficial Soviet efforts to publicize space flight were turning in a more serious direction. Organizations such as OIMS were replaced by laboratories and research organizations dedicated to the design, construction, and testing of rockets. In 1928 the first such organization, founded in Moscow in 1921 to support the black-powder experiments of

Nikolai I. Tikhomirov, was transferred to Leningrad by the Military Revolutionary Council and renamed the Gas Dynamics Laboratory, or GDL. The liquid-propellant branch of the GDL, under the leadership of twenty-one-year-old Valentin Petrovich Glushko, who two years earlier had published an article predicting both space stations and unmanned satellites, conducted important experiments with new propellant combinations and pioneered innovative designs for pumps, exhaust nozzles, and combustion chambers.

Rocket experimentation in Moscow was conducted by the Group for the Study of Reaction Motion, or MosGIRD, established in 1931 under the leadership of Tsander and a young engineer named Sergei Pavlovich Korolev. A similar organization, LenGIRD, was soon operating in Leningrad. Over the next few years, GIRD groups, most of which seem to have had a loose affiliation with MosGIRD, would spring up in Kharkov, Tiflis, Archangel, and other Soviet cities. By 1932 LenGIRD officials reported more than four hundred members; at its peak, MosGIRD and the organizations associated with it claimed as many as one thousand members.

Originally civilian clubs, the GIRD organizations were soon taken under the proverbial wing of the state-controlled Osoaviakhim, the Society for the Promotion of Defense and Aero-Chemical Development. Officially a voluntary organization to promote national defense, Osoaviakhim sponsored glider clubs and other youth activities aimed at military preparedness. It was a convenient way to funnel government research funds to the rocketeers. The official rubles flowed into the MosGIRD coffers in small amounts, but the monies were enough to enable a handful of dedicated young engineers to devote their full time to rocket work.

Soviet rocket organizations made rapid strides during the 1930s. On 17 August 1932 MosGIRD experimenters sent a mixed solid-liquid rocket dubbed the GIRD-09 roaring aloft. Three months later, the same group launched the liquid-propellant GIRD-X, which produced 150 pounds of thrust (667 newtons). Soon thereafter the group was sending liquid-propellant rockets to altitudes of up to thirteen hundred feet (four hundred meters). By 1936 Glushko's branch at the GDL had developed the ORM-65, a rocket engine burning nitric acid and kerosene and producing 385 pounds of thrust.

After 1930 the Stalinist bureaucracy consumed the rocket-research organizations one by one. Most of them were taken over as research teams comprising the Rocket Research Institute (RNII), which was then under the leadership of legendary Red Army commander Marshal Mikhail N. Tukhachevsky. Direct military sponsorship brought with it increased funding and a higher priority, but there was a price to be paid for visibility. The "golden age" of Soviet rocketry came to a sudden end with the execution of Marshal Tukhachevsky during the Great Purge of 1937–38. Anyone associated with Tukhachevsky was suspect. Many simply disappeared. A few, including Sergei Korolev, survived, with great difficulty, to see better days.

Korolev is a name to remember. Born at Zhitomir in the Ukraine on 30 December 1906, he attended high school in Odessa, then moved on to the Polytechnic Institute in Kiev and the Moscow Higher Technical College, where he earned a degree in aeronautical engineering in 1929. He began gliding and designing gliders in his late teens and by 1929–30 was heavily involved in the activities of the group that would become MosGIRD. Over the next seven years he emerged as one of the leading figures in Soviet rocketry.

The Soviet Ministry of Defense published Korolev's first book on rocketry in 1934. In addition to his involvement with the pioneering Soviet liquid-propellant rockets, he took a special interest in the rocket propulsion for aircraft. Working with Glushko, he was involved in the design, construction, and testing of the first two Soviet rocket-propelled aircraft.

Korolev had risen rapidly on the basis of his own talents. As a protégé of Marshal Tukhachevsky, he fell even more rapidly. The young engineer was arrested in June 1938, apparently on the basis of the testimony of his colleague and rival, Valentin Glushko, who had himself been arrested earlier that spring and sentenced to eight years at hard labor. Snatched from his family in the middle of the night, Korolev was accused of "subversion" and of having mismanaged RNII funds. There was no trial. He was beaten until he confessed, then sentenced to a term among the faceless millions in Stalin's Gulag.

After five months of hard labor in the infamous Kolyma gold mines, Korolev was rescued by his old professor of aeronautical engineering, the great aircraft designer A. N. Tupolev, who was supervising the development of new weapons systems at the Tupolevskaya sharaga, a prison for scientists and engineers. Korolev was officially released from custody in 1944 and, ironically, assigned to serve as deputy director of an experimental design bureau headed by another former prisoner, Valentin Glushko, whose testimony had led to Korolev's arrest six years earlier.

Like Sergei Korolev, Wernher Magnus Maxmillian Freiherr von Braun

On 25 November 1932 members of the Group for the Study of
Reaction Motion, or MosGIRD, launched the liquid-propellant
GIRD-X rocket from the Nakhibino Forest near Moscow. Sergei
Korolev, the leader of these smiling enthusiasts, stands at far
left. (Courtesy of the National Air and Space Museum, Smith-
sonian Institution, Photo No. 73-7133)

would spend the years of World War II in the service of the state—but under
very different circumstances. Born at Wirsitz, Posen (then part of Germany,
now in the Polish province of Poznan), on 23 March 1912, von Braun was the
second son of a family that could trace its roots in the landed aristocracy back
to the thirteenth century. His father, Baron Magnus von Braun, represented
the last gasp of conservative Junker tradition. A veteran of the First Potsdam

Guards Regiment, where he had served with Kaiser Wilhelm's sons, he was a confirmed monarchist with a deep-seated contempt for "brutish democracy." Forced out of government during the early years of the Weimar Republic, he pursued a successful career in business and finance. By 1932 the baron's reactionary views were back in fashion. He served as minister of agriculture in the disastrous right-wing cabinet formed by Chancellor Franz von Papen, helping to sound the death knell of the Weimar Republic.

Baroness Emmy Quistorp von Braun approached the business of raising her three sons with the same dedication and seriousness of purpose that her husband applied to high finance and national politics. If she taught her sons that they were members of a privileged elite, and instilled in them the attitudes and values of the old nobility, she also encouraged them to stretch their imaginations and challenged them to match her own extraordinary breadth of interest in science, languages, and the arts.

Young Wernher presented special problems. Enrolled in Berlin's prestigious French Gymnasium, the future engineer did well enough in music and languages, but he was a dismal failure in mathematics and physics. A less imaginative mother might simply have demanded a greater effort from her youngster. Emmy von Braun gave her thirteen-year-old son a telescope. A new and consuming interest in the heavens, along with a transfer to a boarding school at Ettersburg Castle, near Weimar, where formal classwork was combined with hands-on experience in farming, woodworking, and masonry, made all the difference.

Von Braun's grades steadily improved at Ettersburg. The trend continued following his transfer to an allied school on the island of Spiekeroog, in the North Sea, where he led a team of students in the design and construction of a small observatory to house a five-inch telescope.

The final step in the academic transformation of Wernher von Braun began in 1925, when he ran across an advertisement for Hermann Oberth's *Rakete* while thumbing through an astronomy magazine. Fascinated by a book suggesting that one might travel to the distant worlds he had seen through his telescope, von Braun ordered the volume. Discovering that the pages were filled with equations rather than words, he attacked mathematics with a new resolve. Within a few months he was tutoring his fellow students, sending skyrockets crashing through a greenhouse near his parents' home, and, by the late 1920s, careening down Berlin's Tiergartenstrasse in a wagon propelled by six of the largest fireworks rockets money could buy.

Wernher von Braun was not alone in his newfound enthusiasm for space flight. The publication of Oberth's *Rakete* in 1923 had sparked a flurry of interest in the subject. Having borrowed money from his wife to pay the costs of publication, Oberth had the satisfaction of ordering a second printing in 1925. Newspapers chronicled the ongoing debate between Oberth and his academic critics and supporters. Like Goddard, he soon was deluged with letters from admirers, most of whom could not understand his dense mathematical presentation.

The publication of *Die Rakete* inspired a run of other books on space flight. In a brilliant 1925 study of orbital mechanics, architect Walter Hohmann suggested using the gravity of the Moon and planets to sling a spacecraft around the solar system in the most energy-efficient manner, a technique employed by modern space-mission planners. Hermann Potočnik, a former Austrian army captain who wrote under the name Hermann Noordung, produced the first detailed engineering treatise on space stations in 1929, the same year Hermann Oberth took the opportunity to restate his basic ideas, offer up some new thoughts, and attack his critics in the pages of his second book, *Wege zur Raumschiffahrt* (Ways to space flight).

The general public seldom takes much interest in the ideas presented in serious engineering texts, but these pioneering studies of space flight were an exception. "I want to write a feature article about you people," one bewildered newsman remarked to space enthusiast Willy Ley, "and you have written that Hohmann's contribution is important. All right, I believe you . . . *but what does it all mean?*"[5]

Ley and a host of other writers, including Max Valier and Felix Linke, were happy to oblige, producing half a dozen popular treatments of space flight between 1924 and 1931. The brisk sale of these books, and the increasing popularity of science fiction tales by authors such as Otto Willi Gail, offered proof that the space-flight movement had taken root in Germany.

The wave of interest in space travel that swept over the Weimar Republic between 1924 and 1932 was an expression of the Zeitgeist, the spirit of the times. Bled dry and defeated in a catastrophic war, the German people had struggled with revolutionary upheaval, foreign occupation, political uncertainty, and financial collapse. If things were improving a bit by 1924, this was still the age of Dada, psychoanalysis, relativity, the uncertainty principle, and expressionism.

The notion of flying off into space had a certain appeal for the citizens of a

The Soviets, convinced they were the wave of the future, encouraged interest in space flight and published an extraordinarily large numbers of books on the subject, including this 1935 reissue of a Tsiolkovsky volume. (Courtesy of the National Air and Space Museum, Smithsonian Institution, Photo No. A-4320)

nation perched on the edge of chaos. Those who sought to restore German strength and dignity could take pride in Hermann Oberth's contributions to the field of rocketry and support rocket research as a natural extension of Germany's traditional leadership in the development of advanced technologies. The more imaginative and idealistic among them could find hope in the vision of a streamlined, rocket-propelled, utopian future that would include an ultimate escape from both the chains of gravity and the burdens of the past.

However important the cultural context, German interest in space flight was sparked and sustained by the activities of a small band of dedicated publicists and rocket experimenters who were determined to inspire others with their dream. The most influential and colorful of the lot was Max Valier (1895–1930).

An amateur stargazer since childhood, Valier had returned from service with the Austrian air force intending to prepare himself for a career in astronomy. Instead, he became a convert to the "glacial cosmogony" theories of engineer Hans Horbiger, who believed, in a nutshell, that the universe was driven by a "cosmic polarity" between the "glowing substances" that make up stars and the "icy substances" of other celestial bodies. In January 1924, convinced that the only way to prove the theory was to fly into space, Valier purchased a copy of Oberth's *Rakete.* Impressed with Oberth's work, but certain that a treatise filled with equations would never inspire the public or the politicians to support rocket research, Valier arranged a meeting with Oberth and began a book for general readers, *Advance into Outer Space,* which was published late in 1924. The volume was a great success and went through five printings by 1930, when Valier completed a new edition entitled *Raketenfahrt* (rocket travel).

Valier became a familiar figure on the lecture circuit and produced a steady stream of articles on space flight for newspapers and mass-circulation magazines. Ultimately, however, talking and writing about space flight were not enough for him: he was determined to play a personal role in achieving the dream. Enlisting the support of Friedrich Wilhelm Sander, a manufacturer of black-powder rockets, and Fritz von Opel, heir to Germany's leading automobile manufacturer, Valier built and operated a series of railcars, racing automobiles, gliders, bicycles, and sleds powered by black-powder life-saving rockets during 1928 and 1929. These tests did not represent any sort of technological breakthrough, but the sight of a rocket-propelled racing car roaring down the track at 125 miles (200 kilometers) per hour or of a rocket glider flashing across the sky riveted the public's imagination.

The Valier trials propelled German enthusiasm for space flight to a new level. One journal announced the arrival of the *Raketenrummel,* or the "rocket craze," and newspapers and magazines stepped up their coverage of the subject. Toy manufacturers produced miniature rocket cars and planes, and song writers contributed tunes such as "Raketenflug-Marsch." Advertisers were quick to jump on the rocket-propelled bandwagon. "If new goals in space entice researchers," one enterprising copy writer suggested, "to the smoker only one thing is worth striving for: Josetti Juno, Berlin's most celebrated 4 Pf. cigarette." The image of one of Max Valier's rocket planes blasting off into space became a symbol of progress and the hope for a better tomorrow.

The premiere of director Fritz Lang's film *Frau im Mond* (Woman in the Moon) on 15 October 1929 was the high-water mark of the *Raketenrummel.* Weak on plot but strong on special effects, the film offered moviegoers an opportunity to sample the sights, sounds, and the excitement of a trip to the Moon. Lang spent almost as much effort on publicizing the film as he did on its production. He hired Willy Ley, a popular-science writer and rocket enthusiast, to head the publicity effort. Hermann Oberth served as chief technical consultant. As part of his publicity campaign, Lang provided Oberth with a small amount of money for the construction of a liquid-propellant rocket, to be launched in conjunction with the film's premiere. As Goddard had yet to announce the success of his experiments, the stunt could be touted as the world's first flight of a liquid-propellant rocket.

Never much of a success as a hands-on engineer, Oberth struggled to complete the rocket with the help of two assistants. The premiere of the film came and went, and still the three men labored without success. Oberth went home to Romania in disgust but returned in the spring of 1930 to resume the work. Finally, on 23 July 1930, his liquid-oxygen-and-gasoline engine, nicknamed the Kegelduse, or "cone-jet," produced 15.4 pounds of thrust (69 newtons) for ninety seconds during a static test.

Max Valier had also turned his attention to liquid-propellant engines. His work, which was aimed at the development of a power plant for a new rocket car, enjoyed the support of Paul Heylandt, the proprietor of a liquid-oxygen processing plant. On 19 April 1930 Valier and his assistants successfully

A scene from Fritz Lang's *Frau im Mond* (Woman in the Moon), for which Hermann Oberth served as technical advisor. The movie reflected the enthusiasm of Weimar Germans for the new technology as the country was swept by *Der Raketenrummel* (the rocket craze). (Courtesy of the National Air and Space Museum, Smithsonian Institution, Photo No. A-3542)

tested their engine. A photo from this period shows Valier—incredibly—standing only a few feet from the roaring engine. He died a month later, on 17 May 1930, when the same engine exploded.

In the end the honor of flying the first liquid-propellant rocket in Europe went to Johannes Winkler, a graduate engineer, church administrator, and rocket enthusiast who launched his liquid-oxygen-and-methane-propelled HX-1 to an altitude of some two thousand feet (six hundred meters) over Dessau, Germany, on 14 March 1931. Winkler already was a well-known figure in the field, having been elected the first president of the Verein für Raumschiffahrt (Society for Space Travel) and the first editor of its journal, *Die Rakete* (The rocket), the first German serial publication devoted to rocketry and space technology. The VfR, as the society would be known, had been organized by a group of German rocket pioneers meeting in a Breslau alehouse on 5 July 1927. Valier, who had spearheaded the movement to establish the

Paul Heylandt *(left),* **the sponsor, and Max Valier, the designer, stand beside their new rocket car powered by a liquid-propellant rocket engine in 1930. Valier was killed in the explosion of just such an engine on 17 May of the same year. (Courtesy of the National Air and Space Museum, Smithsonian Institution, Photo No. 82-14772)**

VfR as a means of funding Oberth's rocket experiments, declined the presidency because of his lecture and writing schedule.

The VfR was not the first rocket society established in a German-speaking nation. Franz von Hoefft and Baron Guido von Pirquet had organized a similar group in Vienna a year earlier. It was the German rocketeers, however, who would make history. By 1929, the year the organization ceased publishing *Die Rakete* in order to conserve funds and concentrate on experiments with liquid-propellant rockets, the VfR boasted one thousand members.

One of those members was eighteen-year-old Wernher von Braun. Having graduated from the Spiekeroog School, he had moved on to Berlin's Charlottenburg Institute of Technology, where, in the time-honored tradition of a German engineering education, he attended classes and served an apprenticeship as a machinist at the Borsig locomotive works. As busy as he must have been, von Braun joined the VfR in 1929 and volunteered to assist with any contemplated rocket experiments.

As an emerging authority in the field, Willy Ley received a great many letters and visits from young would-be rocketeers, but he never forgot his first meeting with young Wernher von Braun. When the broad-shouldered aristocrat with chiseled features and wavy hair first appeared on their doorstep, Ley's sisters thought that he was almost too handsome, remarking to their brother that the visitor resembled Lord Alfred Douglas, whose friendship with Oscar Wilde had scandalized Europe a generation before. "His manners were as perfect as could be produced by a rigid upbringing," Ley recalled, "and I remember he spoke rather good French. One day he came in while I was struggling with a Sarabande by Handel; after I had finished he sat down and played Beethoven's Moonlight Sonata from memory." An important part of von Braun's education had been music lessons with composer Paul Hindemith.[6]

Ley introduced von Braun to Oberth, who put him to work with Rudolph Nebel, an enthusiast who had assisted with the *Frau im Mond* project, and two other VfR experimenters, Klaus Riedel and Rolf Angel. One of von Braun's first duties was to help set up a display of the unsuccessful but impressive-looking *Frau im Mond* rocket in Berlin's Potsdamer Place and in the basement of the nearby Wertheim Department Store for the celebration of aviation week, 25–31 May 1930. Von Braun, Reidel, and Nebel conducted the successful run of the Kegelduse two months later, shortly after which Oberth again returned to his teaching duties in Romania. That summer and fall, Reidel, Nebel, and a third experimenter, Kurt Heinisch, built and tested the first official VfR project, Mirak 1 ("minimum rocket 1"), on a farm in Saxony that was owned by Riedel's grandparents. The rocket vanished in a burst of flame just after ignition.

Nebel, the senior member of the group, aimed to consolidate his position by establishing a permanent facility for the rocketeers. Col. Karl Becker, a U.S. Army Ordnance officer who had helped to fund work on the Kegelduse, arranged for the VfR to lease an abandoned army ammunition dump and garrison in the Berlin suburb of Reinickendorf. Dubbing the place Raketenflugplatz ("rocket flying field" or "rocketport"), Nebel and the core members of the VfR experimental group moved into the site's abandoned buildings and lived like technological monks, devoted to the cause of space flight.

"The rocket airdrome consisted of a few starkly simple barracks and many workshops," recalled Dimitri Marianoff, Albert Einstein's stepson-in-law. "The impression you took away with you was the frenzied devotion of Nebel's men to their work. Most of them were former officers living under

Rudolph Nebel *(left)* and Klaus Riedel, members of the German rocket society
Verein für Raumschiffahrt, inspect a rocket motor at Reinickendorf, Berlin,
c. 1930. Reidel was killed in a 1944 automobile crash near Peenemünde. (Courtesy
of the National Air and Space Museum, Smithsonian Institution, Photo No. A-3826)

military discipline. Later, I learned that he and his staff lived like hermits. Not one of these men was married, none of them smoked or drank. They belonged exclusively to a world dominated by one single wholehearted idea."[7]

The rocketeers must have been on their best behavior when Marianoff was visiting, for von Braun retained fond memories of raucous drinking parties in neighborhood beer halls. In later years, Nebel portrayed the Raketenflugplatz as a hotbed of idealism and claimed that the VfR team had cooperated with Albert Einstein and other leading scientists to create PANTERRA, an organization designed to "stimulate the peoples of the Earth in the major problems of science and technology, and divert funds for armaments to peaceful and productive work"—in other words, space flight.

There is evidence that Einstein corresponded with some members of the VfR, but it is difficult to imagine what the liberal Zionist might have found in common with Rudolph Nebel, leading light of the Raketenflugplatz. A member of the reactionary veterans' organization Stahlhelm ("steel helmet"), Nebel was already attempting to curry favor with Nazi Party leaders in Munich and Berlin, and soon he would take to wearing a swastika armband at rocket launches.

Brash and arrogant, Rudolph Nebel created the Raketenflugplatz, held it together with the force of his personality, and sustained it with his innate gift for self-promotion. He talked the proprietors of machine shops and factories into contributing materials and labor to rocket research, cadged free meals from the Siemens lunchroom for his unemployed rocketeers, arranged for free gasoline from Shell Oil, and persuaded industrialist Hugo Huckel to provide additional financial support.

Nebel seldom missed an opportunity for publicity. Noting that Henry Ford was visiting Berlin, Nebel sent a telegram to the man who, for Europeans as much as for Americans, was the living symbol of modernism and the power of technology: "I offer the first liquid rocket for the Form Museum. I invite you to inspect the first rocket aerodrome in Berlin Reinickindorf." Ford apparently did not respond, but Nebel later noted that this was the first time he had referred to the test site as the Raketenflugplatz. One can only wonder what Robert Goddard would have had to say about the offer of "the first liquid rocket" to the Ford Museum.[8]

For a little over a year, from March 1931 to April 1932, Nebel and the VfR enthusiasts ensconced at the Raketenfluplatz maintained a high level of activity. According to Willy Ley's accounting, they conducted 270 static-

The young Wernher von Braun *(right)* shown with his friend Constantine D. J. Generales in 1931. The two young men, to the dismay of their landlady, pursued their interest in space flight by testing white mice to destruction aboard a bicycle-wheel centrifuge. (Courtesy of the National Air and Space Museum, Smithsonian Institution, Photo No. 82-11945)

engine tests, 87 flights, 23 demonstrations for other organizations, and 9 presentations for the press. Their rockets reached altitudes up to 4,922 feet (1,552 meters), and in the fall of 1931 one of these missiles came crashing back to Earth through the roof of a building across the street from a police station. The result of this mishap was a "severe reprimand" from a "corpulent police chief" and a series of new safety regulations.

Concentrating on his studies, von Braun remained a peripheral member of the VfR team, but he lost none of his enthusiasm. During the spring and summer of 1931 he attended classes at the Eidgenossische Technische Hochschule in Zurich. There a medical student named Constantine D. J. Generales Jr. made fun of the tall, blond fellow whose "one-track mind" was so focused on flying to the Moon—until von Braun produced a letter from Albert Einstein that discussed space flight in serious terms. Generales quickly converted to the cause and cooperated with von Braun in subjecting white mice to rides on a bicycle-wheel centrifuge designed to test the rodents' capacity to withstand the gravitational forces of space flight. The budding scientists' landlady was understandably appalled to discover a ring of mouse blood around the walls of the room.

After touring Greece with Generales in the summer of 1931 Wernher von Braun returned to Charlottenburg, where he received his degree in engineering in 1932. He planned to begin work on a doctorate in physics at the University of Berlin in the fall; in the meantime he would devote the summer of 1932 to research at Rakentenflugplatz. He soon became curious about the identity of three men who paid repeated visits to the site. Officers of Section 1 (Ballistics and Munitions), Army Ordnance Department, they came dressed in civilian clothes to avoid calling attention to themselves. The officers had noticed von Braun as well. "I had been struck . . . by the energy and shrewdness with which this tall, fair, young student with the broad, massive chin went to work, and by his astonishing theoretical knowledge," Capt. Walter Dornberger recalled many years later. By the end of that summer von Braun and Dornberger had cemented an extraordinary partnership that would have untold consequences for the subsequent history of the twentieth century.[9]

4 • An American Dreamer

The activities of the Verein für Raumschiffahrt at Raketenflugplatz captured the attention of young idealists and dreamers far beyond the borders of Germany. G. Edward Pendray, a reporter for the *New York Herald Tribune,* summed up the matter in an address to the members of the American Interplanetary Society (AIS) on the evening of 13 November 1931. "Whatever the reason," he remarked, "it appears true that the Raketenflugplatz is today the well of inspiration from which all rocket experimenters . . . have drawn."[1]

Nearly two years earlier, on 4 April 1930, Pendray and ten other space enthusiasts had gathered on the third floor of a New York brownstone to found the AIS. Most of them shared an association with an extraordinary man, Hugo Gernsback (1884–1967). Gernsback had emigrated to the United States from Luxembourg in 1904 and had established his first magazine, *Modern Electrics,* four years later. An avid reader of H. G. Wells and Jules Verne, he published his first science fiction story, *Ralph 124C 41,* in 1911–12 as a serial in his own magazine.[2]

The American tradition of pulp science fiction was now under way. Gaudy covers and wildly exciting stories of interplanetary adventure became the hallmarks of such Gernsback publications as *Amazing Stories* and *Science Wonder Stories.* Older Americans might scoff at the notion of space flight, but their sons and daughters devoured pulp "scientifiction" magazines, followed the adventures of Buck Rogers and Flash Gordon in the Sunday funnies, flocked to theaters where they could watch the evil Emperor Ming get his weekly cosmic comeuppance, and proudly displayed the Buck Rogers decoder rings they had received as radio premiums.

Most AIS founders were members of Gernsback's stable of writers, but they took space flight much more seriously than their tales of adventure in the outer reaches of the cosmos might indicate. Initially they were content to plant the seed of enthusiasm for space flight in U.S. soil, sponsoring major lectures on space travel at New York's American Museum of Natural History by such well-known personalities as arctic explorer Sir Hubert Wilkins and arranging the first U.S. showing of Fritz Lang's *Frau im Mond*.

The turning point in the history of the American Interplanetary Society came early in 1931, when Pendray and his wife honeymooned in Germany and toured the VfR's Raketenflugplatz with Willy Ley. Conversation was difficult, as Pendray, the newly elected president of the AIS, spoke no German and the members of the VfR little English. After his return Pendray reported in the AIS *Bulletin* that he had witnessed static tests of the world's first liquid-propellant rocket.

Robert Hutchings Goddard hastened to set the record straight. The American rocket pioneer had accepted a membership in the AIS but refused all invitations to lecture the group on his experiments. As a result, none of the members were aware that their own countryman had built and flown the world's first liquid-propellant rocket five years earlier. They knew only what Goddard had told them in the 1919 Smithsonian pamphlet and what little could be gleaned from sketchy news reports of his activity since that time.

Anxious to set the record straight without revealing too many details of his work, Goddard wrote to Pendray in May 1931. "It may be said that on July 17, 1929, a trial of the liquid-propelled rocket was made at Worcester, Mass., the device functioning satisfactorily as regards the flow of liquid, the ascent of the rocket, and its rapid motion through the air," he explained. "It might be well to add that my work with rocket motors, using liquids and continuous combustion, dates back to 1920."[3]

Even now Goddard did not tell the whole truth, neglecting to mention the first flight of a liquid-propellant rocket, which occurred on 16 March 1926. Instead he based his claim to priority on a July 1929 flight that had caused a brief flurry of comment in the newspapers. Presumably, he was holding the date of his first launch in reserve to counter any potential European rival who might claim to have achieved an earlier flight once Goddard confirmed the 1929 date. Goddard closed his letter to Pendray with a request that the society correct the misinformation printed in the AIS *Bulletin* and explained why he was reluctant to discuss his experiments in detail. "It happens," he noted, "that so many of my ideas and suggestions have been copied abroad without the acknowledgment usual in scientific circles that I have been forced to take this attitude."

In the end Goddard and the AIS would go their separate ways. On 12 November 1932 a core group of AIS experimenters static-tested the society's first liquid-propellant rocket in an open field near Stockton, New Jersey, and made their first flight the following spring. Four more rockets were built and tested over the next two years.

The society's transition from a group of science fiction writers and fans seeking to publicize space flight to a band of dedicated engineers and mechanics struggling to build and fly rockets was reflected in the adoption of a new name on 6 April 1934. Henceforth, the group would be known as the American Rocket Society, or ARS. The ARS became even more seriously focused after 1935, setting to work on the development of a regeneratively cooled, liquid-propellant motor in which the propellants were circulated around the combustion chamber, cooling the engine and preheating the liquids prior to combustion. By 1938 the ARS team, under the leadership of an introverted young engineer named James Wyld, had achieved their goal. In December 1941 Wyld and three ARS colleagues founded Reaction Motors, Inc. (RMI), the first company established in the United States to produce liquid-propellant rocket motors for sale. Only six years later an RMI engine propelled Capt. Charles Yeager and the Bell X-1 through the sound barrier.

For his part Robert Goddard would come to regard 17 July 1929 as one of the luckiest days of his life. Ironically, things had not gone well that day. Alerted by neighbors who were worried about the noisy, flame-spewing rockets being launched from a quiet Massachusetts field, police and news reporters had arrived just as Goddard and his crew were packing up for the day. Their arrival had set in motion a chain of events that led to a ban on all rocket tests in the Commonwealth of Massachusetts. Through the intercession of Smithsonian secretary Charles G. Abbot, Goddard was permitted to continue his tests at a U.S. Army facility, provided that he launch only after it had rained or when there was snow on the ground. On the afternoon of 22 No-

G. Edward Pendray makes adjustments to the American Rocket Society's first rocket, the ARS-1, in Stockton, New Jersey, c. 1932. Pendray's wife, Leatrice, stands by while Hugh Franklin Pierce, one of the designers, warms his hands. This rocket was never fired. (Courtesy of the G. Edward Pendray Collection, National Air and Space Museum, Smithsonian Institution, Photo No. A-4555)

vember 1929 Goddard was sitting at his desk planning the next expedition to Fort Devens, Massachusetts, his new launch site, when he got a telephone call from Charles Augustus Lindbergh.

Lindbergh had been thinking about rockets for more than two years. Shortly after his triumphant return from France in the summer of 1927, he had set out on a nationwide aerial tour designed to encourage local airport construction. The experience of flying in and out of countless substandard airfields convinced him of the need for an emergency rocket that could lift an airplane to safety in the event of an engine failure on takeoff.

On 1 November 1929 Lindbergh flew to Wilmington, Delaware, to discuss the possibility of such a device with Du Pont engineers. They advised against the idea, arguing that a rocket of the sort he described would have to be charged with 400 pounds (180 kilograms) of powder and would require a combustion chamber lined with firebrick. Several weeks later, a friend called Lindbergh's attention to an article in the latest issue of *Popular Science Monthly* detailing Goddard's work and outlining his recent problems with the Massachusetts authorities. It was, Lindbergh would recall, "one of those unpredictable incidents . . . that so often bend the trends of life and history."[4]

Lindbergh drove to Worcester for his first meeting with Goddard on 23 November 1929. By the end of his visit the notion of a rocket that could lift airplanes out of danger had been replaced by a much grander vision. "I am sure Professor Goddard had no idea how his words set my mind to spinning," Lindbergh wrote in his autobiography. "Flight to the moon theoretically possible! . . . Then space was to be an extension of, not a limit to, the works of man. The rocket, like the wheel, the hull, and the wing, would throw back old horizons. . . . Maybe man would learn how to travel faster than the speed of light. Impossible? Who dared, now, to say anything was impossible!"[5]

Impressed with the potential of rocketry, Lindbergh offered to help obtain additional funding to move things along. "This struck a responsive chord in me," Goddard reported to his friend Charles Abbot at the Smithsonian. "The rocket developments being made in Germany, apparently very substantially supported, make me impatient when things do not move as fast as possible."[6] When Lindbergh asked how much money would be required to make a real difference, Goddard responded that $100,000 would enable him to accomplish in four years what might otherwise take a lifetime. "Driving back to New York that evening," Lindbergh recalled, "I considered ways in which I might help Goddard obtain one hundred thousand dollars. It was almost seven times as much as I had raised to finance, just three years earlier, the *Spirit of St. Louis* and the flight to Paris."[7]

Over the next few weeks, Lindbergh arranged for another meeting between Goddard and Du Pont officials, encouraged John C. Merriam of the Carnegie Institution of Washington, D.C., to supplement the existing Smithsonian grant, and spoke to his financial contacts in New York and California. Finally, he presented the case to the "the man who combined the wealth, vision, and courage . . . necessary to provide the adequate financing of Professor Goddard—Daniel Guggenheim."[8]

The multimillionaire son of a Swiss immigrant who had made his fortune in copper mining, Guggenheim and his son Harry, a World War I aviator, took a special interest in the promotion of aeronautics. After a ten-minute conversation, during which Lindbergh described the promise of Goddard's work and the extent to which German researchers were forging ahead, Daniel Guggenheim agreed to contribute an initial fifty thousand dollars to a special Clark University account that would be designated the Daniel Guggenheim Fund for Measurement and Investigation of High Altitudes. Before the project was over, Guggenheim would spend some $188,500 on Goddard's work.

Guggenheim's largess enabled Goddard to devote his full attention to rocketry. The scientist's first step was to find the most suitable area in which to test his creations. It would have to be isolated, located far from the prying eyes of the press and of curious neighbors, and characterized by clear skies and flat, open terrain. Presented with those requirements, a meteorologist suggested the high-plateau country around Roswell, in southeastern New Mexico. The Goddards drove into the remote town on 25 July 1930, fell in love with the place, and promptly rented Mescalero Ranch.

Goddard would spend the eight happiest years of his life—from 1930 to 1932 and from 1934 to 1942—living in Roswell. On leave from his teaching duties at Clark, he labored in the machine shop behind the lovely adobe house, designing, building, testing, and rebuilding the components of his rockets. In the process he developed gyroscopic control systems, regeneratively cooled motors, parachute-recovery systems, and variable-thrust rockets. He was one of those people who are never satisfied. There was always a new refinement—an improved pump, a better gasket, a tighter screw. And he wanted to do all the work himself. He was never more content than he was when laboring over a rocket with a wrench or a screwdriver in hand.

While in New Mexico, Goddard produced three major types of rockets, those known as the A, K, and L series. The largest of the L series stood 18 feet, 5.75 inches (5.63 meters) in height, weighed 170 pounds (77.3 kilograms) fully fueled, and developed more than 470 pounds of thrust (2,091 newtons). By 1935 L-series rockets had flown faster than the speed of sound and had reached altitudes up to 7,700 feet (2,350 meters). It was as fast, and as high, as any Goddard rocket would ever fly, but the rockets did not achieve such performances with any predictability. Lindbergh and Harry Guggenheim made several trips to witness a flight, but as luck would have it, something went wrong every time.

Isolated and self-contained, Roswell boasted a well-known boarding school and a small arts colony. Goddard himself joined the Rotary Club and, at the urging of a new friend, artist Peter Hurd, took up painting. The physics professor instantly became one of Roswell's most distinguished citizens, and the ranchers, merchants, artists, and teachers who were his friends and neighbors respected his work, if they did not fully understand it.

In the desert vastness Goddard could insulate himself from the pressures and perceived dangers of the outside world. As far as his rockets were concerned, he trusted no one outside his tiny circle of assistants, each of whom was either a longtime friend or a member of the family. He recorded his progress in his own notebooks and in the confidential technical reports he submitted to the Guggenheims, the Smithsonian, and Clark University. Rocketeers all over the world knew his name, but none of them knew for certain what he had accomplished.

By the mid-1930s Goddard sensed that the time had come to preserve a record of his achievement. True to his nature, however, he approached that task gingerly. After considerable urging on the part of Charles Lindbergh and Harry Guggenheim, he ordered his workers to prepare one of the successful K-series rockets for presentation to the Smithsonian. Museum officials were informed that the donation was intended only to establish the state of Goddard's art; under no circumstances was it to be shown to anyone. As an afterthought, Goddard allowed that Smithsonian personnel at least could unscrew the top of the crate and *look* at the rocket if they wished. In the end the device was consigned to a storage closet in the basement of the institution's Castle building on the Mall, where it would remain, unseen and unexamined, for the next thirteen years.

The Smithsonian published Goddard's second major paper, *Liquid-Propellant Rocket Development,* in the spring of 1936. Again, however, un-

**Robert Goddard at his easel in Roswell, New Mexico, c. 1932.
The artist Peter Hurd, another resident of this high plateau
town, encouraged Goddard to take up painting. (Courtesy of
the National Air and Space Museum, Smithsonian Institution,
Photo No. 98-16193)**

like most leading scientists of his generation, who could point with pride to a
long list of articles in professional journals, Goddard published very little,
and much of that was in mass-circulation magazines such as *Popular Science
Monthly* or *Scientific American.* He spent far more time filing patent documents and preparing detailed affidavits that established his priority in key areas of research. All told he would receive some 214 patents covering virtually
every phase of liquid-propellant rocketry.

Goddard's second Smithsonian paper provided him with yet another
means of ensuring his position of leadership in the field of rocketry. Although

At the launch site near Roswell, New Mexico, on 23 September 1935. *Left to right:* A. W. Kisk, Goddard's in-law and employee; Harry Guggenheim; Robert Goddard; Charles Lindbergh; and N. T. Ljungquist, a Goddard workman. (Courtesy of the National Air and Space Museum, Smithsonian Institution, Photo No. A-4968)

the new pamphlet was even shorter than *A Method of Reaching Extreme Altitudes* and contained few technical details of any value to other rocket builders, it offered the first summary of his work since 1919. Once again he was caught in a flurry of press interest, including a "Time Marches On" radio broadcast about his work.

At the Guggenheim Aeronautical Laboratory, California Institute of Technology (GALCIT), a small group of graduate students and rocket enthusiasts took a special interest in Goddard's new report. Under the leadership of Hungarian-born scientist and engineer Theodore von Kármán, GALCIT had become a world center for the research and teaching of aerodynamics, fluid mechanics, and structures. Under von Kármán's guidance, a small group of graduate students, led by a young engineer named Frank Malina, were also pursuing rocket research.

Goddard first learned of this group in August 1936, when he visited Robert

Millikan, president of the California Institute of Technology, or Caltech. As the head of a university hosting a Guggenheim-funded laboratory and as a member of the informal board established by the Guggenheims to advise on Goddard's research, Millikan hoped to forge an alliance between the GALCIT team and the senior American space theorist and rocket builder. His assumption that Goddard would jump at the opportunity to work with von Kármán's brilliant young students, however, was incorrect.

Going through his mail on the day of his return to Roswell, Goddard found a letter from Millikan alerting him to the fact that Frank Malina, whom Goddard had met in Pasadena, was planning to stop at Roswell during a visit to his home in Texas. Goddard was surprised but hospitable. He picked the young man up at the train station that afternoon, took him to dinner, and saw him safely back to the hotel that night. The following morning he drove Malina out to the launch tower and engine-test stand and then brought him back to the house for a talk.

Goddard was sufficiently impressed with the young engineer to suggest that Malina consider coming to work for him. Hospitality had its limits, however. The older man listened to Malina's description of the GALCIT project and offered his advice, but he made certain that every piece of rocket equipment was covered with a drop cloth before his guest set foot inside the workshop.

"I wish I could have been of greater assistance to him," Goddard wrote to Millikan after Malina's departure, "but it happens that the subject of his work, namely the development of an oxygen-gasoline rocket motor, has been one of the chief problems of my own research work, and I naturally cannot turn over the results of many years of investigation, still incomplete, for use as a student's thesis."[9]

"The . . . impression I obtained," Malina recalled many years later, "was that he felt that rockets were his private preserve, so that any others working on them took on the aspect of intruders. He did not appear to realize that in other countries were men who, independently of him, . . . had arrived at the same basic ideas for rocket propulsion. His attitude caused him to turn his back on the scientific tradition of communication of results through established scientific journals, and instead he spent much of his time on patents."[10]

Goddard left Roswell in 1941 to work with U.S. Navy and Curtiss-Wright engineers on the development of jet-assisted-takeoff and variable-thrust,

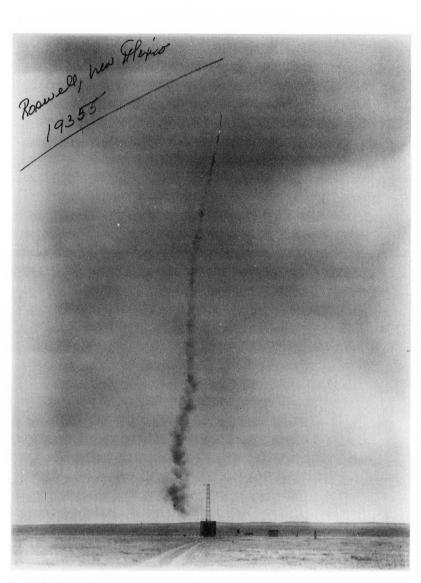

Launch of a Goddard A-series rocket, Roswell, New Mexico, 1935. Although Lindbergh and Guggenheim were never fortunate enough to see it, rockets of this type achieved altitudes of more than seven thousand feet. (Courtesy of the National Air and Space Museum, Smithsonian Institution, Photo No. A-1074)

liquid-propellant rockets. By the spring of 1944 he was receiving detailed reports on a new German long-range rocket, the V-2. "The weapon is reported to be almost identical with the rocket we were working on in New Mexico at the time we changed over to war work," he wrote to Harry Guggenheim, "except that it is larger."[11]

Goddard provided the editor of the *National Geographic News Bulletin* with a list of his own patents for almost every aspect of V-2 design. "So closely do the mechanical features of V-2 parallel the American projectile [Goddard's rocket]," the *News Bulletin* announced in January 1945, "that some physicists think the Germans may have actually copied most of the design."[12]

That certainly was the opinion of Robert Hutchings Goddard. On 14 August 1945 he died of throat cancer, convinced that his work had played a key role in the Germans' success. It simply was not true. The Germans had followed the same path taken by Goddard, quite unaware that he had been there before them. Under the inspired leadership of Wernher von Braun, they had surged past him without a backward glance, achieving Goddard's goal of sending a rocket to the edge of space.

"There is no direct line from Goddard to present-day rocketry," Frank Malina's mentor, the redoubtable Theodore von Kármán, commented. "He is on a branch that died. He was an inventive man and had a good scientific foundation, but he was not a creator of science, and he took himself too seriously. If he had taken others into his confidence, I think he would have developed workable high-altitude rockets and his achievements would have been greater than they were. But not listening to, or communicating with, other qualified people hindered his accomplishments."[13]

In terms of the cold, hard facts that serve as the measure of an engineering program, von Kármán was correct. It is difficult to believe that Robert Goddard could not have come closer to his goal had he been able and willing to build a team of the sort that Wernher von Braun worked to create. As Charles Lindbergh recognized, however, there are other ways to gauge the achievements of an individual human spirit. "When I see a rocket rising from its pad," Lindbergh wrote in 1974, "I think of how the most fantastic dreams come true, of how dreams have formed into matter, and matter into dreams. Then I sense Goddard standing at my side, his human physical substance now ethereal, his dreams substantive. When I watched the fantastic launching of

Apollo 8, carrying its three astronauts on man's first voyage to the moon, I thought about how the launching of a dream can be more fantastic still, for the material product of dreams are limited in a way dreams themselves are not. What sunbound astronaut's experience can equal that of Robert Goddard, whose body stayed on earth while he voyaged through the galaxies?"[14]

5 • *Vergeltungswaffe!*

Learned Faustus,
To find the secrets of astronomy
Graven in the book of Jove's high firmament,
Did mount him up to scale Olympus' top:
Where, sitting in a chariot burning bright
Drawn by the strength of yoked dragons' necks
He views the clouds, the planets, and the stars,
The tropics, zones, and quarters of the sky,
From the bright circle of the horned moon
Even to the height of the *primum mobile.*
Christopher Marlowe, *Doctor Faustus*

On 5 May 1945 Maj. Anatoli Vavilov led elements of Gen. Konstantin Rokossovsky's Second White Russian Army through the gates of Peenemünde, a sprawling research complex on the northern tip of Usedom Island in the Baltic Sea. The officers of the unit had been briefed thoroughly for this operation. They carried lists of German scientists and engineers who were to be located and held and photographs of Wernher von Braun, a man in whom their superiors were particularly interested. With the exception of a handful of stragglers, however, Peenemünde was abandoned. The Soviets had captured the ground, but the talent, key items of equipment, and much of the technical data were nowhere to be found.

Three days earlier PFC Frederick P. Schneikert, an American rifleman with an antitank company of the 324th Infantry Regiment, Forty-fourth Division, was guarding the approaches to the Austrian border town of Schattwald when a well-dressed civilian came peddling down the road on a bicycle. The man braked to a stop when challenged and introduced himself as Magnus von Braun. In an excited mixture of English and German he explained that his brother, the inventor of the V-2, was hiding in the woods nearby with a group of scientists and engineers, all of whom were anxious "to see Ike as soon as possible."[1]

The story of German interest in the military potential of the liquid-propellant rocket begins with a talented artillery officer named Karl Emil Becker. A ballistics expert with a doctorate in engineering, Becker had been involved with the great Paris Gun of 1918, which had fired waist-high projectiles at Paris from a distance of 80 miles (129 kilometers). Considering the matter in the postwar years, he came to believe that any additional increase in range, accuracy, or payload would require a radically different technology.

In 1926 Becker assisted one of his former professors in revising a *Textbook of Ballistics,* the second volume of which included a discussion of rocketry, complete with a treatment of the space-flight proposals that had been offered by Robert Goddard and Hermann Oberth. Three years later Lieutenant Colonel Becker, now chief of the ballistics and munitions section of German army ordnance, decided to investigate the potential of the rocket as a long-range artillery weapon. The fact that the Versailles treaty, which officially ended World War I, did not specifically forbid Germany to develop rockets has often been cited as a key element in Becker's decision. Given his prior interest in the subject, the excitement generated by the *Raketenrummel,* the growing number of experimental rocket groups springing up around the country, the relatively small amount of initial funding required, and German readiness to pursue weapons research that *was* banned, there can be little doubt that Becker would have pursued rocket research in secret even if it had been prohibited by the treaty.

Within a year Becker had hired four subordinates, each an officer-graduate of the Technische Hochschule Berlin with an advanced degree in mechanical engineering. Together they set out to survey the growing body of literature on rocketry. The first turning point in their work came in December 1930, when army budget officials approved the equivalent of fifty thousand dollars for rocket research and promised an equal amount the following year.

Eventually, the German army rocket program would pursue two goals. First, they would develop a solid-propellant barrage rocket with a range of six miles (seven kilometers); second, the group would seek to create a liquid-propellant weapon with twice the range and payload of the Paris Gun. Tests of existing solid-propellant rockets soon were underway at Kummersdorf, an army ordnance testing ground about twenty-five miles (forty kilometers) south of Berlin.

The liquid-propellant weapon presented a much greater challenge. Initially Becker turned to the experimental rocket groups for expertise. In the summer of 1930, even before his own program was fully funded, he supported Oberth's work on the Kegelduse motor. That fall he provided the members of the Verein für Raumschiffahrt with small amounts of research funding, and he arranged for Rudolph Nebel to lease the land for Raketenflugplatz. Appalled by Nebel's wild promotional schemes and his obvious delight in sensational press coverage, Becker concluded that the crew at Raketenflugplatz was hopelessly ill-suited for secret government work.

Becker and his fellow officers were paying periodic visits to the launch site by the spring of 1932 and were not particularly impressed. The absence of test procedures was particularly troublesome. The VfR seldom measured an engine's rate of fuel consumption or the amount of thrust it generated. In an effort to gather basic engineering data Becker signed two contracts with Paul Heylandt, the liquid-oxygen manufacturer who had supported Max Valier, for the production of a motor that could be static-tested at Kummersdorf.

Finally, anxious to see how one of the VfR rockets would perform under rigid test conditions, the army ordnance staff invited Nebel and his associates to conduct a launch at Kummersdorf. If the rocket, a craft thirteen feet (four meters) long that Willy Ley had dubbed "the two-stick repulsor," rose straight to altitude and deployed a parachute and a flare, the group would be paid. If not, the rocketeers would bear the expenses of the day.

Early one morning in June 1932 Nebel, Klaus Riedel, and Wernher von Braun drove through the gates of Kummersdorf with the rocket and launch rail strapped on their automobile. The launched rocket climbed into the clouds hanging low overhead, arched over, and crashed to earth less than a mile away without deploying the parachute or flare. Nebel spent the next few weeks trying to get the ordnance office to pay the bills in spite of the agreement. Becker refused to budge, promising only that Nebel and the VfR might be given subcontracts for the components of future rockets, which would be constructed "by competent officers."

The entire ordnance staff agreed with Capt. Walter Dornberger's conclusion that "rocketry was a sphere of activity beset with humbugs, charlatans, and scientific cranks, and sparsely populated with men of real ability."[2] The officers were determined to abandon the civilian experimenters in favor of an in-house rocket-development program. When the twenty-year-old von Braun visited Becker's office to plead the VfR case, he was sent back to the Raketenflugplatz with a proposition: the army would hire any member of the group willing to work on a systematic research program conducted behind the fence of an army post.

The VfR response was less than overwhelming. Nebel complained that government red tape and bureaucracy would only impede the work, and most of the others agreed. After von Braun signed his first contract with army ordnance in late October 1932 he took with him only one veteran of the Raketenflugplatz, machinist Heinrich Gruenow. Alone among the rocketeers, von Braun recognized that only the government could afford to support the cost of developing large rockets. "Our feelings toward the Army resembled those of the early aviation pioneers, who . . . tried to milk the military purse for their own ends and who felt little moral scruples as to the possible future use of their brainchild," he explained. "The issue . . . was simply how the golden cow could be milked most successfully."[3]

Weapons research was only a means to an end for the young von Braun. In developing a long-range rocket weapon for the army he also would be moving toward the realization of his own dream of space flight. "When the die was cast, " he pointed out many years later, "the Nazis were not yet in power, and to all of us Hitler was just another mountebank on the political stage."[4] Willy Ley would remember von Braun as a young man with little interest in politics. "In von Braun's opinion (as of that time), the German Republic was no good and the Nazis were ridiculous."[5]

On 30 January 1933, however, just three months after von Braun had signed his first contract with Colonel Becker, the "mountebank" and his "ridiculous" cohorts came to power in Germany. As the nature of the Nazi regime became clear, there is no indication that von Braun so much as blinked. One golden cow was as good as another. Like Johann Faust, an earlier German scientist of note, von Braun had struck a pact with demons.

The design and construction of a rocket that would serve both as a test vehicle for the ordnance department and as the subject of von Braun's doctoral thesis moved slowly at first. Funds were limited, and von Braun still had a great deal to learn, about both engineering and the fine art of getting things done in a government bureaucracy. By the summer of 1933, however, he had begun construction of a rocket that would be known as Aggregat 1 ("unit" 1), or the A-1.

It was a small rocket, standing just 4 feet, 7 inches (1.4 meters) tall and weighing only 330 pounds (150 kilograms). The engine burned liquid oxygen and ethyl alcohol and developed 660 pounds of thrust (2,940 newtons). An electric motor spun a weight in the nose prior to launch to provide gyroscopic stability. A learning experience fraught with difficulties, the project climaxed in the destruction of the little craft during a test run. "It took us exactly one-half year to build," von Braun quipped, "and exactly one-half second to blow it up."[6] The accident did not destroy the confidence of the ordnance department, however. Becker believed in von Braun and accepted setbacks as part of the price of eventual success. With his own program underway, Becker now took steps to either bring the other rocketeers under his wing or drive them from the field. He first turned his attention to the remnants of the Raketenflugplatz crowd, whose penchant for publicity continued to attract unwanted attention to the rocket work.

Rudolph Nebel had remained true to form after von Braun's departure, accepting a contract from the city of Magdeburg to launch a manned rocket into the upper atmosphere as part of a local celebration scheduled for Pentecost 1933. The large rocket was never constructed, but on 29 June 1933 a preliminary version fell out of the rack and went skittering across the ground with its engine running full blast.

The wild goings-on in Magdeburg created a fatal split between Nebel and the more stable members of the VfR. Willy Ley and Major Helmuth von Dickhuth-Harrach, president of the group, charged Nebel with fraud. Becker applied a more direct remedy, arranging a Gestapo raid on the Raketenflugplatz in the late spring of 1933. The records of the organization were seized, and many of its talented technicians were either brought into von Braun's program or provided with jobs in industry.

Nebel's attempts to interest the air ministry and the leaders of the Nazi paramilitary organization Sturmabteilung (SA) in establishing rocket programs were to no avail. As for von Braun, he could not imagine Nebel's functioning successfully on the team of disciplined engineers he was building, but he never lost his fondness for the scoundrel who had guided his own early involvement in rocketry. Thus he arranged for Nebel to receive a sum of money for a worthless patent and, in 1943, involved his consulting firm in a wartime facilities-construction program.

Von Braun was the primary beneficiary of Becker's crackdown on civilian rocket experiments. By the end of 1934 Walter Riedel and Arthur Rudolph, both veterans of the Valier/Heylandt efforts, had been folded into his team. Klaus Riedel (no relation to Walter), a mainstay of the VfR who had remained a Nebel loyalist to the bitter end, finally joined the von Braun team in 1937. In 1933, after Becker was promoted to general, D'Aubigny Engelbrunner Ritter von Horstig and Walter Dornberger, both original members of the ordnance rocket group, took over military command of the program.

By June 1934 von Braun was completing his doctorate in physics at Friedrich-Wilhelm University and pressing ahead with work on the A-2 rocket. The new craft was about the same size as its predecessor, but it featured a number of improvements. As had been the case with the A-1, the engine, designed to run for sixteen seconds, was surrounded by the ethyl-alcohol tank for more efficient cooling, and the spinning stabilizer weight was repositioned to the midpoint of the vehicle, between the propellant tanks. In December 1934 von Braun and his crew transported a pair of the new A-2 rockets to Borkum Island in the North Sea. Nicknamed Max and Moritz after the two obstreperous youngsters known as Hans and Fritz in the American version of the German comic strip *Katzenjammer Kids,* the rockets successfully climbed away from the launcher and ascended to an altitude of 6,500 feet (2,000 meters). The engines and stabilizers functioned perfectly. Von Braun had proven that his team could build rockets—and that rockets were potentially worth building.

Back at Kummersdorf, work soon was underway on plans for the A-3. The new missile would represent a quantum leap in size, standing 25 feet (7.6 meters) tall, weighing five times as much as the A-2, and featuring an engine producing 3,300 pounds of thrust (14,680 newtons) for forty seconds. The A-3 also featured the most advanced guidance-and-control system ever flown on a rocket: five gyroscopes and two accelerometers would sense deviations from a preset course and actuate four molybdenum steering vanes set in the exhaust.

The program had reached a critical juncture. Engine development and static testing could be conducted at Kummersdorf, but the facility was too small for flight testing even the A-2. By 1936 eighty workers were reporting to von Braun; that number would continue to grow as the program entered a new phase. Von Braun already had begun to recruit a new group of specialists from industry and the universities, including engineer Bernhard Tessmann and chemist Walter Thiel, who eventually would take command of engine development. Clearly, the rocket project had outgrown Kummersdorf.

It was Emmy von Braun, the engineer's mother, who suggested the site for an expanded facility. She had grown up near the Pomeranian town of Anklam, a few dozen kilometers south of the point where the Peene River entered the Baltic Sea. Her father had hunted ducks near Peenemünde, a small fishing village on the island of Usedom at the mouth of the Peene. Unlike the popular Baltic resorts on the nearby island of Rügen, Peenemünde had remained undeveloped. The area was an isolated hunter's paradise of marshlands, dunes, pine forest, and long, sandy beaches—a perfect natural missile test range stretching far down the Baltic coast.

Initial funding for a development facility on Peenemünde quickly fell into place. For some time, Maj. Wolfram von Richthofen, a cousin of the Red Baron air ace of World War I, had been interested in the potential of rocket weapons and auxiliary-propulsion systems for aircraft. Von Braun, pleased to cooperate, conducted tests for the Luftwaffe and generally encouraged its interest in rocketry. Von Richthofen, for his part, agreed to contribute 5 million reichsmarks—about $1.2 million—toward a joint facility.

Not to be outdone by a Luftwaffe rival, General Becker allocated an additional 6 million reichsmarks, or the equivalent of $2.4 million, to the project. By the spring of 1936, work was underway on something new in the world, a self-contained research-and-development center, complete with all the facilities required to design, build, and launch large rockets. When complete, Peenemünde would house two distinct programs. Peenemünde West was a relatively small Luftwaffe facility where the Fieseler 103, better known as the V-1, was tested. A winged, pilotless drone powered by an Argus pulsejet, the V-1 first flew in 1942 and was capable of carrying 1,870 pounds (850 kilograms) of high explosives over a range of 150 miles (240 kilometers).

Peenemünde East was von Braun's domain. This facility cost an estimated

**Wernher von Braun (*center,* in the dark suit) with German offi-
cers at Peenemünde, 1943. Von Braun headed the massive
rocket development center built near this isolated fishing village
on the Baltic Sea island of Usedom, where his grandfather had
hunted wild fowl. (Courtesy of the National Air and Space Museum, Smithsonian Institution, Photo No. 78-5935)**

300 million reichsmarks (more than $70 million) to construct. At its peak in
1943 it employed more than 1,950 scientists, engineers, and technicians.
Thousands of other workers, including specialists in industry, academia, and
the military, were also involved in the program. The center's total budget for
1943 reached 112 million reichsmarks, or almost $27 million.

Peenemünde was far more than a series of launch pads and test stands strung along the Baltic coast. It featured a liquid-oxygen plant, well-equipped machine shops, and the most advanced supersonic wind tunnels in the world. There were facilities for materials testing and for the development of telemetry, guidance, and control equipment, and one factory for the construction of A-4 missiles. There were headquarters buildings, warehouses, assembly buildings, and a power plant that supplied the electrical needs of the entire facility. Housing for the scientists and engineers, barracks for the troops, and a labor camp for "foreign workers" were located in the nearby village of Karlshagen. A prisoner-of-war camp at Trassenheide provided additional forced labor for both army and Luftwaffe operations.

On 4 December 1937 von Braun and his crew launched an A-3 from the Greifswalder Oie, a tiny island in the Baltic within sight of Usedom. It was the first of some seventy launches that would be conducted from this remote facility prior to the completion of the main launch pads at Peenemünde in 1941. The A-3 itself was a failure, but it helped pave the way to success. When all four test vehicles came to grief as a result of guidance-system failures, von Braun's team went to work on an A-5 model. First launched from the Oie in October 1938, the A-5 reached altitudes of up to eight miles (thirteen kilometers).

All of this effort was but a prelude to the real business of Peenemünde—the design, construction, testing, and launch of the von Braun team's masterpiece, the A-4. The big rocket was trundled out of the shops at Peenemünde for the first time in the spring of 1942. Poised on the launch stand, it stood 46.1 feet (14.06 meters) tall, measured 65 inches (165 centimeters) in diameter, and weighed more than 28,000 pounds (12,727 kilograms). It had only one purpose: to drop one ton (910 kilograms) of high explosive on targets as far as 220 miles (350 kilometers) away from the launch point. The A-4 would make the trip in five minutes, roaring up to a maximum altitude of 60 miles (96 kilometers) above the earth, then rushing down toward its target at 3,500 miles (5,600 kilometers) per hour, 4.4 times the speed of sound.

Every feature of the A-4 was the result of a prodigious research-and-development effort. The engine, which delivered 59,500 pounds of thrust (264,700 newtons) for a full sixty-eight seconds, was the last in a series of rocket motors designed, built, and tested at Kummersdorf and Peenemünde. The combination of hydrogen peroxide and potassium permanganate generated the high-pressure steam that drove centrifugal pumps, which in turn fed alcohol and liquid oxygen from the tanks to the combustion chamber at a rate of more than 50 gallons (190 liters) per second. The combustion chamber itself was a masterpiece of engineering, as was the guidance-and-control system and virtually every other element of the rocket.

The creativity required to produce a technological marvel like the A-4 was not restricted to the workshops and laboratories. Wernher von Braun led the way at every step. He wrote the specifications for the various systems, chose the team leaders, contributed ideas, and put all of the pieces together. With the rocket design complete, he turned to the problems of testing and manufacturing. Thousands of people contributed to the final product, but von Braun was that rarest of commodities: an indispensable man. Together with Becker, von Horstig, and Dornberger, von Braun also fought the political battles, including the constant struggle to maintain funding and to fend off, in so far as possible, the repeated attempts of Heinrich Himmler's SS (or Schutzstaffel, the special Nazi security force) to take full control of the V-2 program from the army. Von Braun and his colleagues were not always victorious, but they did keep the project moving forward.

The first V-2 was destroyed in an explosion during a ground test on 18 March 1942. An A-4 left the launch pad for the first time on 13 June. It tumbled out of the low-hanging clouds ninety seconds later and crashed into the Baltic less than a mile from shore. A second launch, on 16 August, ended in an explosion a little more than seven miles (eleven kilometers) up.

Von Braun may have sensed that the third time would be the charm. He ordered a technician to decorate the tail of the next rocket with a talisman—a small drawing of a young woman jauntily perched on the lower horn of a crescent Moon. It was a reminder of the *Frau im Mond* film project that had marked von Braun's initiation into the fraternity of rocketeers more than a decade earlier. On 3 October 1942 an A-4 finally performed according to plan. As fate would have it, Hermann Oberth, the man who had inspired them all, was on hand at Peenemünde that day. "As he shook hands and congratulated me," Dornberger recalled, "I could only say that the day on which we had been privileged to take the first step into space must also be a day of success and rejoicing for him, and that congratulations should go to him for showing us the way." Underscoring the underlying dream of space flight, von Braun quipped in later years that there was only one problem with his rockets—they hit the wrong planet.[7]

Three more rockets were launched by the end of 1942, but none of them matched the success of the October flight. Things had improved a bit by the spring of 1943, but when Dornberger and von Braun flew to East Prussia to brief Hitler on the program, the film of the October launch was still their most impressive exhibit. It was enough. The rocket project was given the highest priority, and Joseph Goebbels, the minister of propaganda, provided the missile with a new name: V-2, for *Vergeltungswaffe,* or "vengeance weapon," number two.

While the engineers pressed forward with the research and development effort, plans for the production and operational use of the missiles already were underway. Klaus Riedel was placed in charge of designing a mobile launching system for the rocket, and Arthur Rudolph became chief engineer of Peenemünde's pilot production plant. Von Braun had argued against the in-house manufacture of operational rockets. He saw Peenemünde as a laboratory and testing ground, not as a factory. Dornberger and the army thought otherwise.

Albert Speer, Hitler's architect and armaments minister, placed Gerhard

A test version of the V-2 mounted on a mobile launcher of the sort used to send operational missiles on their way to enemy targets. (Courtesy of the National Air and Space Museum, Smithsonian Institution, Photo No. 76-2003)

Degenkolb, who had earned a reputation for ruthless adherence to schedules in the construction of locomotives, in charge of a special committee to oversee the mass production of the A-4 rockets. The actual assembly of the missiles was to be carried out at three sites: the Peenemünde pilot production plant, the Rax-werke at Wiener Neudstadt, Austria; and the Zeppelin works in Friedrichshafen.

Arthur Rudolph and his colleagues faced overwhelming difficulties in developing a production system that could be employed in all three rocket factories. The A-4 was an evolving piece of technology, not a finished product. An estimated sixty-five thousand alterations, large and small, were made to the rocket between the time of its first flight and the operational deployment of the first missiles twenty-three months later. Each of the approximately 250 missiles produced by the Peenemünde development center emerged as a hand-crafted original.

The activity at Peenemünde had not gone unnoticed in London. As early as 1939, reports of a German long-range rocket program had filtered back to British intelligence; by June 1943 Peenemünde had been identified as the principal German rocket center. Between 11:25 P.M. and 3:40 A.M. on the night of 17–18 August 1943, 596 aircraft of the Royal Air Force (RAF) Bomber Command dropped eighteen hundred tons of bombs on Peenemünde. Code-named Project Hydra, the attack took the lives of some 780 individuals on the ground: 180 Germans, including propulsion expert Walter Thiel, and between 500 and 600 foreign laborers, who were housed in flimsy shacks with no bomb shelter.

The attack did not destroy Peenemünde, but it did force the decision to separate research and development from production. The manufacture of the V-2 now shifted to a set of tunnels under Kohnstein Mountain, near Niedersachswerfen in the Harz Mountains. Begun in 1917 as gypsum and anhydrite mines, the tunnels had been extended between the years 1935 and 1943 and used to store gasoline and chemical weapons.

Further extensions and additions transformed the tunnels into a V-weapons production site. When work was completed, the facility consisted of two central tunnels, each 2 miles (3.2 kilometers) long, that cut across a series of forty-six work tunnels, each of which measured 500 feet (152 meters) long, 30 feet (9 meters) wide, and up to 22 feet (6.7 meters) high. Although some German civilians were employed in constructing the tunnel factory, the bulk of the work was performed by slave laborers. Initially housed in the tunnels, the slave laborers were eventually moved into Dora, a concentration camp that would continue to house the workers once V-weapons fabrication began.

The facility was named for the new, government-controlled company Mittelwerk G.m.b.H. (Central Works Limited), which oversaw rocket production. The entire operation was supervised by a board that included representatives of the SS, which supplied labor from Dora. At best the arrangement was nothing more than a makeshift organization through which all of the factions competing for control of the V-2 production effort could achieve a minimum level of cooperation.

Finally, in early 1945, SS Reichsführer Heinrich Himmler and his underling Hans Kammler took full command of the program. Historian David Irving estimates that some 6,089 V-2s were produced between 1942 and 1945. Approximately 300 missiles were expended in test firings, and some 2,550 were captured by the Allied armies that eventually overran Mittelwerk and V-2 launch units in the field. The remainder of the rockets were committed to operational use.

The vengeance-weapon campaign began shortly before 4:30 A.M. on 13 June 1944, when the first V-1 fell on Swanscomb, near London. The first V-2 to strike an enemy target fell near the Porte d'Italie in Paris on the morning of 8 September. The V-2 assault on London began that evening. Between that date and the time of the last operational V-2 firing on 29 March 1945, some 3,400 rockets were prepared for launch against Allied targets. One hundred and seventy of those missiles were lost in accidents at launch or early in flight. Allied sources reported the detonation of roughly 2,890 missiles. Presumably the missing V-2s broke up in the air or fell into the sea.

Historian Michael Neufeld conservatively estimates the total cost of the German guided-missile program at RM 2 billion, or roughly half a billion U.S. World War II dollars. That figure is one-quarter of the $2 billion cost of the U.S. WW II atomic-weapons program. Given the disparity in the strength of the two economies, it is clear that vengeance weapons came with an astronomical price tag.[8]

There is a wide range of opinions as to the V-2's effectiveness. Gen. Dwight D. Eisenhower, supreme commander of the Allied Expeditionary Forces in Europe, thought that the Normandy invasion might have been "difficult or impossible" if the vengeance weapons had been targeted on troop concentrations in southern England during the winter and early spring of

The main entrance to Mittelwerk, the elaborate underground V-2 factory near Niedersachswerfen, Germany, in the Harz Mountains. (Courtesy of the National Air and Space Museum, Smithsonian Institution, Photo No. 79-13171)

1944. Winston Churchill disagreed, as did Nazi arms minister Albert Speer, who pointed out that a V-2, which was destroyed at the conclusion of a single mission, cost almost as much as a fighter aircraft. "The rockets were a very expensive affair for us," Speer maintained, "and their effect compared to the cost of their output was negligible."[9]

Postwar studies have revealed that V-1 and V-2 attacks on England, Belgium, and France destroyed some 33,700 buildings and damaged another 204,000. The vengeance weapons killed 12,685 people and wounded another 26,433. The V-2 campaign against London alone took 2,700 lives, an average of 1.76 persons for each of the 1,528 rockets that rained down on the English capital. In contrast, as many as 130,000 persons died as a direct result of the single atomic bomb dropped on Hiroshima, Japan.

But a simple accounting of the dead masks the real horror of the vengeance weapons. The V-2 was unique in the history of modern warfare, in that more people died building the rockets than lost their lives on the receiving end. Between 1943 and 1945, 60,000 slave laborers passed through the gates of Dora and other concentration camps associated with Mittelwerk. One-third of them—20,000 human beings—died there. There is no accounting of the additional thousands of slave laborers who perished at the liquid-oxygen plants and other facilities related to V-2 operations.

When Arthur Rudolph arrived in September 1943 to take command of technical matters related to rocket manufacture, Dora had yet to be constructed. At that point, laborers worked, ate, slept, and died in the tunnels they were constructing. "We worked in the center of the mountain with no air, and just had one small piece of bread and margarine to eat all day," a member of the French resistance recalled. "It was horrible."[10] The laborers came from all over occupied Europe: Russia, Poland, Hungary, and France. Disease, exhaustion, exposure, and malnutrition took the heaviest toll. Initially bodies were trucked to the Buchenwald extermination camp for disposal. As the number of dead began to climb, however, Dora was provided with its own crematorium.

Punishment for suspected sabotage at Mittelwerk was swift and certain. One former prisoner remembered one hanging that was conducted from an overhead crane near Arthur Rudolph's office. Twelve prisoners were lined up beneath the crane. Their hands were bound and sticks were tied in their mouths with wire to prevent them from screaming. At a signal, the crane slowly lifted them above the heads of the assembled work force. "Instead of letting them drop and killing them on the spot," he recalled, "they let them hang very slowly with pain that's absolutely horrible." The bodies were left hanging for some time as an object lesson to other workers.[11]

Conditions at Mittelwerk deteriorated rapidly during the final four months of its operation. Determined to wring the last ounce of work from the prisoners, officials increased the beatings and intensified the search for saboteurs and malingerers. Rations were reduced or cut off entirely to punish those who failed to achieve rising production quotas. "They had been starved to death," one of the American liberators said of the prisoners. "Their arms were just little sticks, their legs had practically no flesh on them at all."[12]

Charles Lindbergh, who had worked so hard on Robert Goddard's behalf a decade earlier, toured Mittelwerk soon after its liberation. He was driven through the tunnels in a jeep, past the abandoned bits and pieces of V-2 rockets that would never fly, and then back into the open for a visit to Dora. As he stood before the crematory ovens, still filled with ash and charred bone,

Slave labor for the Mittelwerk V-2 complex came from all over occupied Europe.
Of the sixty thousand laborers sent there, twenty thousand died. (Courtesy of the
National Air and Space Museum, Smithsonian Institution, Photo No. 94-3375)

Lindbergh was approached by a skeletal looking young man dressed in the
striped clothes of a prisoner. Holding his cupped hands before him, the survivor explained: "Twenty-five thousand in a year and a half, and from each
there is only so much."[13]

On the day Charles Lindbergh was introduced to the ghastly reality of
Dora, Wernher von Braun and his colleagues were already safe in American
hands. Few of the engineers had experienced the horrors of Mittelwerk firsthand, and none of those who had been involved with the underground factory
expressed any sense of responsibility for what had transpired there. The horror had been the work of the SS, they were quick to explain. The engineers
had been building weapons for the defense of their nation, and their underlying goal had always been the conquest of space.

As weapons developers the Germans were little different from the technical professionals in Allied countries. Although they may express occasional
moral compunctions, the scientists and engineers of every nation are nor-

mally expected to pursue weapons research as a patriotic duty. Moreover, there is a wealth of evidence indicating that space flight was never far from the minds of the leading members of the rocket team.

On the evening of 3 October 1942, in celebration of the first successful launch of a V-2, Walter Dornberger hosted a party at the Peenemünde officers' club. "The following points may be deemed of decisive significance in the history of technology," he told his guests. "We have invaded space with our rockets for the first time—mark this well—have used space as a bridge between two points on earth; we have proved rocket propulsion practical for space travel. To land, sea, and air may now be added infinite space as a medium of future intercontinental travel."[14]

Seven years earlier Arthur Rudolph had shared quarters with von Braun in the mess hall at Kummersdorf. "We didn't like to get up early," he recalled. "We liked to work late at night instead . . . . At midnight von Braun had his best ideas, and his ideas led to one thing: space travel." Indeed, it was during those late-night sessions in the mid-1930s that von Braun first began to work out the details of a manned voyage to Mars, a project to which he would return time and again over the next two decades.[15] In February 1938 the residents of Peenemünde had celebrated "Mardi Gras on Mars," complete with a Martian goddess who welcomed Wernher von Braun to her "planet." By 1940 von Braun had set the engineers in his advanced-projects office to work on the design of A-4 follow-on vehicles that would be capable of either dropping a payload of high explosives on New York City or orbiting an Earth satellite. "We didn't want to build weapons," Rudolph stated, "we wanted to go into space. Building weapons was just a stepping stone, the only one available."[16]

The Nazi overseers of the rocket team were fully aware of the strong commitment to the ultimate goal of space travel among the leaders of the group. Wernher von Braun was arrested by the Gestapo in 1944 and accused of placing the goal of space flight above the military objectives of the Reich. In fact, the engineer had become a pawn in the bureaucratic struggle for control of the rocket program. General Dornberger secured his release two weeks later by pointing out that von Braun was indispensable to the program. The incident was proof of just how dangerous life could be for the highly visible manager of a well-funded and very promising program operating in a hazardous political environment dominated by the most ruthless figures of the regime—men such as Heinrich Himmler and Hans Kammler.

A historian writing fifty years after the fact, with no personal experience of life in a regime in which terror is a governing principal, must exercise great caution in commenting on the behavior and the decisions of individuals who passed through that fire. At the same time, respect for the thousands of prisoners who suffered and died building wonder weapons in the tunnels beneath Kohnstein Mountain demands that we recognize just what was involved in milking the golden cow of the Third Reich.

Take the case of Arthur Rudolph. There is no reason to doubt the depth of Rudolph's commitment to the dream of space flight, but it is important to note that he seems to have been fully committed to the regime for which he was building weapons. He had joined the Nazi Party in 1931 and enlisted in the SA reserve two years later. When questioned about his involvement with those organizations by representatives of the U.S. Department of Justice half a century later, Rudolph explained that he had joined the party to express opposition to communism and to support "the preservation of western culture."[17] "I read *Mein Kampf,* and agreed with a lot of things in it," Rudolph commented to a reporter forty years later. "Hitler's first six years, until the war started, were really marvelous. They were the best years in Germany. Everybody was happy. Everybody got jobs."[18] Those were also, of course, the years of terror, repression, censorship, and the systematic exclusion of Jews from national life.

No member of the rocket team was charged with a war crime. In the mid-1980s, however, Arthur Rudolph voluntarily gave up his U.S. citizenship and returned to Germany to avoid denaturalization proceedings by the Immigration and Naturalization Service. "In this case we have not just the utilization of slave labor, which is a war crime under the Nuremburg laws," explained the Department of Justice official handling Rudolph's case, [but] "we have the outright exploitation of slave labor working under terrible conditions." German officials who considered the matter did not charge Rudolph with war crimes. His repeated requests for permission to return to the United States, or to Canada, prior to his death, however, were rejected.

Wernher von Braun's situation was very different and much more complex. He did not join the Nazis until 1937, when the membership rolls were reopened to admit those potential leaders who had not been blessed with Arthur Rudolph's foresight. In 1940, when Heinrich Himmler offered von Braun a commission in the SS, the young engineer polled his colleagues for their opinion before accepting. He received three pro forma promotions in the SS, ending the war as a major.

In later life von Braun would do his best to conceal his association with the SS. When a U.S. Army general asked him about the truth of columnist Drew Pearson's assertion that he had been a member of that organization, von Braun admitted that it was true but asked that the information be held in confidence, "as any publicity would harm my work with NASA." Clearly, however, his association with the SS, like his membership in the Nazi Party, was the result of political opportunism, not ideological conviction.[19]

The rocket engineers did not create the concentration camps, but they certainly recognized that slave labor was to be the means of achieving the goals they had persuaded Nazi leaders to pursue. Moreover, it was Arthur Rudolph, not the SS, who first suggested the use of forced labor at the Peenemünde production facility.[20] Certainly the handful of engineers who were heavily involved in the rocket production program saw the slave labor system at its worst. Von Braun himself made twelve to fifteen visits to the Mittelwerk facility. Rudolph, who headed the production office at Nordhausen, witnessed hangings and other punishments.

The operation of the slave labor system was the responsibility of the SS, but the rocket engineers were regarded as members of the team. Von Braun, Dornberger, and Rudolph attended a meeting at Nordhausen on 6 May 1944 at which a Mittelwerk official announced the solution to a V-2 production bottleneck: The SS would be asked to round up an additional eighteen hundred skilled slave laborers from occupied France. It obviously would have been foolhardy for von Braun, only a month out of a Gestapo prison cell, to raise an objection at that point. At the same time, the meeting is a clear indication of the extent to which the engineers were kept informed of matters relating to slave labor.

Ultimately it is impossible to escape the realization that if the Nazis made use of the rocket team, the opposite also was true. The regime could have spent the vast sums of money lavished on the V-2 on any number of other, potentially more effective weapons. The engineers, together with their allies in ordnance, not only created the V-2 but also worked very hard to sell it to an initially skeptical government. The members of the German rocket team, and the U.S. and Soviet governments that would profit from their knowledge and experience, were the ultimate beneficiaries of the V-2 program.

The V-2 was a truly revolutionary piece of technology, exemplifying the complexities, ironies, ambiguities, and unexpected consequences of large-scale technical and scientific projects sponsored by twentieth-century nation-states. A brilliant engineering and managerial achievement, and a symbol of Nazi scientific and technical capability, it was nonetheless a miserable failure in its primary role as a weapons system. Captured rockets that had been hand-crafted by slave laborers working in the tunnels they had carved beneath the surface of Kohnstein Mountain ended their careers carrying scientific instruments to the edge of space high over the New Mexican desert, not all that far from where Robert Goddard had launched his rockets a decade or so earlier. A failed weapons system, the product of horrific slave labor, the V-2 had opened the door to the universe.

6 • "Our Germans"

Well, that was a crying need in my new job; rescue those able and intelligent Jerrys from behind the barbed wire, and get them going in our military projects. . . . Feed them into American industry. . . . It may come as a shock to some readers to learn that Werhner von Braun was one of those people, and also his old teacher and boss, General Dornberger. Think of that, and it really makes you sit up and take notice. Wonder where we'd be today if we'd let those people languish in the pen? *General Curtis E. LeMay, 1965*

From 1945, when the Russians captured all the German scientists at Peenemünde[,] . . . they have centered all their attention on the ballistic missile. *President Dwight D. Eisenhower, 1957*

We captured the wrong Germans! *U.S. Department of Defense spokesperson, 1957*

Like Wernher von Braun, Frank J. Malina spent the war years at the head of a rocket team. The son of Czechoslovakian immigrants who had settled in Bernham, Texas, Malina earned a bachelor of science degree in mechanical engineering from Texas A&M University, then began his graduate work at the Guggenheim Aeronautical Laboratory, California Institute of Technology in 1934. He was a dedicated student and earned master's degrees in both mechanical and aeronautical engineering before pressing ahead with work on a doctorate.

Malina's master's thesis, a conventional study of propeller efficiencies, convinced him of the need for new propulsion systems. In 1935 fellow student William Bollay presented a seminar paper on the rocket research of Austrian scientist Eugen Sanger that fanned Malina's spark of interest into a flame. Not long after Bollay's presentation John Whiteside Parsons and Edward S. Forman, neither of whom were students at the California Institute of Technology, introduced themselves to Malina. Science fiction enthusiasts who had followed the growth of the space-flight movement in Germany with keen interest, they had corresponded with pioneers such as Willy Ley and had begun to build and test their own solid-propellant rockets. Now they were anxious to begin work with liquid-propellant motors, and, recognizing their own limitations, they had come to Caltech for advice and support.

Malina was impressed. Forman, a gifted mechanic and machinist, was a straightforward space-flight enthusiast of the kind that had populated Germany's Raketenflugplatz. Parsons, the brains of this two-man outfit, was much more complex. A dark, handsome man, he had dropped out of the University of Southern California after two years, continuing the study of chemistry on his own. By the time he met Malina he was a twenty-one-year-old expert in explosives who had transformed the basement of his widowed mother's home on South Orange Avenue, Pasadena's "millionaire's row," into a small-scale munitions factory.

Years later Malina recalled that Parsons "loved poetry and the exotic aspects of life."[1] Exotic, indeed! At the time Malina met him, Parsons was a member in good standing of what he would describe to curious Federal Bureau of Investigation (FBI) agents as "the Church of Thelema." In fact, the group was the Los Angeles Agape Chapter of the Ordo Templi Orientis, a satanic cult based on the teachings of the famous English black magician Aleister Crowley. A decade later, struggling to convince federal authorities that he really was not a politically suspicious person, Parsons described the group as "a sort of mythic love cult."[2] That was putting the Church of Thelema in the best possible light. As local wizard-in-chief, Parsons presided over a series of bizarre rites, including a ceremony in which naked women were encouraged to jump through hoops of fire.

As a leader of the witches of greater Pasadena, Parsons would play a significant role in the history of California cults. He met L. Ron Hubbard, a young naval officer and pulp science fiction writer, at a meeting of the Los Angeles Science Fiction and Fantasy Society. The two became fast friends

John Whiteside Parsons *(left)* **and Edward S. Forman** *(right),*
leading members of the California Institute of Technology's
"suicide squad," standing in front of the historic Ercoupe mono-
plane that served as the first flying test bed for the rocket-
assisted takeoff system developed at Caltech's Guggenheim
Aeronautical Laboratory. (Courtesy of the National Air and
Space Museum, Smithsonian Institution, Photo No. 98-16194)

who shared a house and, for a time, a mistress. Within five years Hubbard had published *Dianetics* (1950) and was attracting the first followers to what would become the Church of Scientology. Modern scientologists insist that Hubbard was simply an intelligence officer gathering information on Parsons and his followers.

Armed with the promise of help from Parsons and Forman, Frank Malina decided to pursue his doctoral degree in the study of rockets. His research would consider the thermodynamics of rocket motors, the design of a sounding rocket that was capable of lofting scientific instruments into the upper atmosphere, and the accumulation of static-test data on both solid- and liquid-propellant motors.

Clark B. Millikan, the son of Caltech president Robert A. Millikan and one of the bright young stars of the GALCIT faculty, rejected Malina's proposal, advising him instead to select a topic that would prepare him for a high-paying job in the aircraft industry. Undaunted, Malina turned to Theodore von Kármán, the brilliant Hungarian scientist and engineer whose leadership of GALCIT had put Caltech on the map. Von Kármán, an expert in high-speed aerodynamics who had lectured on rocketry, approved Malina's unorthodox proposal.

By the spring of 1936 two more GALCIT graduate students had joined Malina's team. Aerodynamicist Apollo M. O. Smith would assist in rocket design, and Hsue Shen Tsien, a doctoral candidate whom von Kármán considered his most gifted student, would assist with thermodynamics. Weld Arnold was another welcome addition to the group. A graduate assistant in astrophysics, he offered to contribute one thousand dollars to the project in return for a position as volunteer photographer. Malina jumped at the chance. He and Parsons had become so desperate for funds that they had set to work on a novel, in which they had hoped to interest Hollywood, about a group of rocket engineers. Now Malina could afford to set fiction aside and stop eating at Parsons's house, where exotic and usually explosive chemicals shared the kitchen shelves with canned goods and cereal.

At the time of their abortive contact with rocketeer Robert Goddard in 1936 Malina and his group already had established a small static-test stand for their first gaseous-oxygen-and-methyl-alcohol motor in the Arroyo Seco, near the Devil's Gate Dam on the outskirts of Pasadena. In fact, work began to progress on several fronts simultaneously. While the popping roar of ex-

Relaxing at the site of the original rocket test stand in Arroyo Seco, California, 1936.
Left to right: **Rudolf Schott, Apollo Smith, Frank Malina, Edward Forman, and John Parsons. (Courtesy of the National Air and Space Museum, Smithsonian Institution, Photo No. 75-1149)**

perimental rocket motors echoed through the arroyo, Parsons experimented with efficient but incredibly corrosive propellants.

On one occasion, Malina and Smith accidentally released a cloud of choking gas inside the Caltech chemistry budding. Within hours, a thin layer of bright red rust had coated the steel parts of virtually every piece of exposed equipment in the lab. On another occasion, the two scientists spilled nitrogen tetroxide on the building's lawn. The resulting brown spots lasted for weeks, prompting the groundskeeper to banish members of the GALCIT "suicide squad" from the well-manicured campus lawns.

In August 1938 Reuben Fleet, president of Consolidated Aircraft of California, approached Caltech for advice on the possibility of employing rockets to boost large seaplanes into the air. In the end Consolidated did not act immediately on Malina's recommendations, but from that point on, the notion of rocket-assisted takeoff was in the wind. The use of such technology had occurred to both Charles Lindbergh and to Reuben Fleet; now it occurred to U.S. Army Air Corps general Henry H. "Hap" Arnold, who requested a report on its possibilities from the National Academy of Sciences Committee on Army Air Corps Research.

Von Kármán and Caltech president Millikan arranged for Malina to present a report based on his preliminary research for Consolidated at a December 1938 meeting of this committee. On the basis of Malina's presentation, the academy provided GALCIT with a contract for one thousand dollars to develop a formal proposal for a research program on rocket-assisted takeoff. In July 1939 the academy accepted GALCIT's proposal and awarded the aeronautical laboratory a ten-thousand-dollar grant for the development of liquid- and/or solid-propellant rocket units designed to lift heavily laden planes off the runway.

Jerome Hunsaker of the Massachusetts Institute of Technology (MIT), who had won a similar contract to develop aircraft-windshield deicers, remarked that von Kármán and Malina were welcome to "the Buck Rogers job." Eager to be taken seriously, the Caltech team struck the word "rocket" from their vocabulary. Instead the scientists identified their proposal as "Air Corps Jet Propulsion Research: GALCIT Project No. 1." The Caltech facility that grew out of their research program would be named the Jet Propulsion Laboratory, or JPL, in 1944.[3]

Suddenly, rockets were big business at Caltech. Von Kármán himself took on the role of project director, and Malina served as chief engineer. Parsons concentrated on developing a trustworthy solid-rocket motor with a relatively long burning time; Forman supervised all of the required machining; and a new team member, Martin Summerfield, was placed in charge of developing a suitable liquid-propellant motor.

On 12 August 1941 twelve of Parsons's jet-assisted-takeoff, or JATO, rockets boosted into the air a small Ercoupe monoplane piloted by Capt. Homer A. Boushey Jr. of the U.S. Army Air Force (USAAF). The following April the first flight test of one of Martin Summerfield's liquid-JATO motors was made from the dry lake bed at Muroc, California. Having successfully designed, manufactured, and tested two satisfactory products, the members of the GALCIT team, with the approval of the Caltech administration, set themselves up in business: assisted by Andrew G. Haley, an attorney and a friend of von Kármán, they chartered the Aerojet Engineering Corporation on 19 March 1942 to produce and sell their rockets.

With JATO technology in hand by the summer of 1943, the GALCIT/Aerojet team turned its attention to the long-range artillery rocket. From U.S.

Members of the GALCIT team signing the wing of the Ercoupe, the United States' first rocket-assisted airplane, prior to takeoff, 12 August 1941. *Left to right:* **Clark Millikan, Martin Summerfield, Theodore von Kármán, Frank Malina, and Capt. Homer Boushey Jr., the plane's pilot. (Courtesy of the National Air and Space Museum, Smithsonian Institution, Photo No. 75-10233)**

Army Air Force officials von Kármán and his colleagues received copies of the earliest classified intelligence reports on the V-2, and Malina and Tsien provided an engineering analysis of the weapon's potential. In return, Maj. Gen. Gladeon M. Barnes, chief of the Research and Development Division of U.S. Army Ordnance, offered GALCIT, now reorganized into the Jet Propulsion Laboratory, a contract to undertake the development of a missile capable of delivering a 1,000-pound (450-kilogram) payload over a range of 75 to 100 miles (120 to 161 kilometers). As in the case of the JATO project, the team decided to pursue research on both solid- and liquid-propellant rockets. The solid-propellant vehicles, the first of which was test-fired in early December 1944, would be identified as Private A and Private F. One of the liquid-propellant rockets, Corporal E, featured a pressure-fed motor, and the other, Corporal F, employed turbine-driven pumps.

By the spring of 1945 JPL had more than demonstrated its capacity to meet the demands of wartime research and development. To handle the workload, Frank Malina's original three-man team had grown into an organization that employed 264 scientists, engineers, technicians, and crafts persons, not to mention all the workers at Aerojet General. JPL's projects included JATO units, rocket weapons, ramjets, and guidance and control systems. In addition the laboratory pursued basic research in thermodynamics, high-speed aerodynamics, instrumentation, and a host of related fields.

General Barnes and Col. Gervais William Trichel, head of the newly established rocket branch of the U.S. Army Ordnance Research and Development Division, had every confidence in JPL. Nevertheless, as evidence of the magnitude and complexity of the effort required to produce the V-2 began to emerge, the leaders of U.S. Army Ordnance decided to offer a contract for the study of long-range ballistic missiles to a firm with experience in managing large projects.

Late in 1944 Colonel Trichel authorized General Electric (GE), the company that had produced the first U.S. jet engine, to begin work on Project Hermes, the goal of which was to design and build a missile comparable to the German V-2. General Electric officials were pleased to get the contract but loath to assign the new job to any of their leading turbine specialists or metallurgists, all of whom were busy developing potentially profitable jet engines. As a result, thirty-two-year-old Richard Porter, who held a doctorate from Yale and who had completed work on a sophisticated fire-control system for the Boeing B-29, found himself at the head of the Project Hermes team.

Colonel Trichel felt that the best way to begin this new project was to study examples of German missile technology and talk with the scientists who actually had designed and built the rockets. In February 1945, therefore, he dispatched Maj. Robert B. Staver to Europe with instructions to locate the German rocket engineers and to pave the way for technical debriefings by Dick Porter and his GE colleagues.

Trichel also asked Col. Holger N. Toftoy, chief of U.S. Army Ordnance Technical Intelligence in Europe, to prepare one hundred operational V-2 rockets for shipment to the United States. On 11 April 1945 Colonel Toftoy, a West Point graduate with considerable experience in both artillery and ordnance, received word that U.S. troops had overrun an enormous underground V-2 factory—Mittelwerk—near Nordhausen. Slated to become part of the Soviet zone, the site was scheduled to be turned over to the Red Army within

Frank Malina *(left)* and Martin Summerfield *(center)* work on a GALCIT-designed, Douglas Aircraft–produced WAC Corporal Sounding rocket at the White Sands, New Mexico, test range in the late 1940s. Malina and his team, in this less sensitive era, regarded their scientific rocket as a "little sister" to the larger Corporal E missile and dubbed it the WAC Corporal for the Women's Auxiliary Corps. (Courtesy of the National Air and Space Museum, Smithsonian Institution, Photo No. 84-14740)

eight weeks. With no time to lose, Toftoy immediately dispatched ordnance personnel and transport troops to the area to remove everything from complete rockets to bits and pieces of equipment. Meanwhile, Staver and his small crew scoured the Thuringian countryside in search of the scientists and technicians who had worked on the missiles. He had no way of knowing that Wernher von Braun and several hundred Peenemünde team members were searching just as hard for him, or someone like him. Years later Richard Porter was asked who had been most responsible for bringing the German rocket team to the U.S. "Probably von Braun himself," he responded, "as much as anyone else."[4]

As early as January 1945 von Braun and Walter Dornberger had decided to take the future into their own hands. The war, they realized, was lost, but the danger was far from over. Their Peenemünde team was receiving conflicting orders from a disintegrating command structure and faced the prospect of being overrun by the victorious Red Army or massacred by SS units determined to prevent the secrets of Nazi weaponry from falling into Allied hands.

With these threats to their safety in mind, the leaders of the rocket team orchestrated their escape as carefully as they had planned the development of the V-2. Their goal was clear: to move the members of the rocket team and their families to a safe haven where they could await an opportunity to surrender as a group, striking a bargain that would enable the scientists to continue their work on behalf of another government. There was never any question as to whom they wished to surrender. "We despise the French; we are mortally afraid of the Soviets," one of them explained. "We do not believe the British can afford us; so that leaves the Americans."[5]

The first train left Peenemünde on 17 February 1945 with several hundred engineers and family members packed into boxcars. The next three months were filled with numerous crises, ranging from encounters with the SS to a car crash that left von Braun with his arm and shoulder in a cast. Nevertheless, by Victory in Europe Day, 8 May 1945, nearly five hundred German engineers, along with their wives and children, were the guests of the U.S. Army at a compound near Garmisch-Partenkirchen that originally was built to house athletes at the 1936 Winter Olympics. Allied interrogators, each one anxious to uncover the secrets of the Nazi rocket-weapons program, flocked to the area.

"They didn't know what to ask," Dornberger later recalled. "It was like they were talking Chinese to us." Fritz Zwicky, a Caltech colleague of von

Kármán, noted that the Germans "watched the unexpected and disorderly procedures of the British and American teams with discerning eyes and it became apparent that they considered our missions pretty much of a farce."[6] When Dick Porter arrived at Garmisch in the middle of May he recognized exactly what was going on. "He [von Braun] and his people were really looking to make a bargain with the Yanks, if they could. . . . Some days they'd talk; some days they'd get together and decide 'we're going to hold out for a while and see if we can make a deal with the Yanks for this, that or the other.' Fairly early on, they decided they would like to come to the United States to work."[7]

Also fairly early on, their U.S. Army hosts reached a reciprocal decision. In fact, from the outset ordnance officers on the scene favored the transferal of key German personnel to the United States, where they could continue their work. Von Braun already had pointed out the virtues of such a plan in debriefing documents prepared for the Allies at Garmisch, and he spent a considerable amount of time that spring interviewing Peenemünde team members and briefing Staver on potential recruits for the U.S. program.

On 19 July 1945 the U.S. Joint Chiefs of Staff issued a memorandum announcing the creation of Project Overcast. Under the terms of this program, 350 German scientists and engineers would be offered six-month contracts for defense work in the United States. "Virulent Nazis" and persons who might be charged with war crimes were to be excluded, and in no case were family members to accompany those selected for inclusion in the program. Colonel Toftoy, who recently had replaced Colonel Trichel as chief of the U.S. Army Ordnance Rocket Branch, received 100 of these 350 total contracts for members of the Peenemünde team. Returning to Europe, he consulted with von Braun, Porter, Staver, and others, then took it upon himself to enlarge his quota to 115 men whom he regarded as absolutely essential.

Later in July U.S. officials sent critical members of the Peenemünde staff to Cuxhaven in Germany for a debriefing by British intelligence and a demonstration launch of a few V-2s over the North Sea. Von Braun, Dornberger, and four of their department chiefs flew to London for high-level debriefings. After these meetings von Braun and his subordinates were returned to Germany, but British officials decided to hold Dornberger while they investigated his role in the vengeance-weapons campaign. The general remained in custody until July 1947, at which time he was allowed to accept a research contract with the U.S. Air Force (USAF). Two decades later, Dornberger, who had served as military commander of the world's first strategic-missile program, retired as vice-president of Bell Aircraft.

Meanwhile, on 20 September 1945 von Braun and six of his fellow Peenemünde "prisoners of peace" were ushered through the gates of Fort Strong, a U.S. Army post located on an island in Boston Harbor. His colleagues were sent on to Aberdeen Proving Ground, north of Baltimore, to begin the process of sorting and translating captured documents; von Braun was put on a train for Fort Bliss, Texas, where, over the next four months, the cream of the Peenemünde team would join him.

The idea of establishing what one newspaper referred to as "das Amerikanische Peenemünde" on the West Texas prairie did not sit well in all quarters. Rabbi Stephen Wise, for one, objected to allowing people whom he regarded as "ardent pro-Nazis" into the country while concentration-camp survivors were refused visas. "Red tape, lack of shipping facilities, and every other possible handicap face these oppressed people," Wise commented, "while their oppressors are brought to this country . . . and are favorably housed and supported at our expense."[8]

In Washington, D.C., officials of the Departments of State, Labor, Commerce, and Justice lodged various protests relating to the quality of the background checks that had been run on the Germans, the nature of their entry into the United States, their access to classified information, and the lax security procedures at Fort Bliss. And although von Braun himself was always excepted, the credentials of the other German rocket experts were called into question. One American scientist characterized the Germans as "high-class radio hams"; another commented that "the Germans . . . actually represent men considered second- or third-rate in their fields."[9]

In terms of the actual projects the Germans would be working on, a number of leading American scientists regarded the possibility of an intercontinental ballistic missile (ICBM) as little more than a pipe dream. Vannevar Bush, for example, director of the Pentagon's Office of Scientific Research and Development and chairman of the Joint Chiefs of Staff Committee on New Weapons, expressed annoyance at talk of a rocket capable of accurately delivering nuclear weapons over a range of 3,000 miles (4,830 kilometers). "In my opinion, such a thing is impossible," he remarked, "and will be impossible for many years to come."

In fact, Bush was responding to a wave of interest and enthusiasm for rocketry and space flight that had swept through the Pentagon in 1945–46. Taking

The Paperclip Germans at Fort Bliss, Texas, 1945. Wernher von Braun is in the front row, seventh from the right, with his hand in his pocket. (Courtesy of the National Air and Space Museum, Smithsonian Institution, Photo No. 77-14246)

the other side of the issue in a November 1945 report to the secretary of war, Gen. Henry "Hap" Arnold, wartime commander of the U.S. Army Air Force, not only predicted the advent of nuclear-tipped missiles but also forecast a day in the not-too-distant future when the nations of the world would build "space ships capable of operating outside the atmosphere."[10]

Arnold's bold prediction was not just empty talk. While U.S. Army Ordnance was preparing to fly captured V-2s as a first step toward the development of long-range U.S. rockets, U.S. Navy and Air Force officials were looking even further into the future. The country's first serious consideration of the military potential of space flight surfaced in October 1945, when officials of the U.S. Navy's Bureau of Aeronautics, or BuAer, unveiled a proposal for an experimental program aimed at orbiting an artificial Earth satellite.

The navy's interest in rocketry dated back to 1935–38, when Midshipman Robert Truax began testing experimental liquid-propellant motors at the U.S. Naval Engineering Experiment Station in Annapolis. BuAer planners recognized that a satellite could be put into orbit by adding stages to an existing V-2. With the advice and assistance of GALCIT, Aerojet General, and North American Aviation, they now proposed developing the much more advanced high-altitude test vehicle (HATV), to be powered by a liquid-hydrogen-and-liquid-oxygen main engine and eight solid-propellant motors.

In March 1946, with cost estimates for the satellite project reaching $8 million, BuAer officials invited the U.S. Army Air Force to buy into a joint program. However, Gen. Curtis E. LeMay, deputy chief of staff for USAAF Research and Development, saw no reason why the USAAF should share the conquest of a new frontier with another service, particularly the navy. He rejected the proposal and gave his friends at Project Research and Development (RAND) three weeks to come up with its own satellite plan, demonstrating that the USAAF was just as serious about this space business as was the navy.

Originally a nonprofit subsidiary of Douglas Aircraft, RAND had been created at the request of the USAAF in 1945 to conduct special studies and offer expert advice on the procurement of advanced technologies. When completed, RAND's satellite study offered a choice between a three-stage, liquid-oxygen-and-liquid-hydrogen vehicle or a four-stage, liquid-oxygen-and-alcohol-propelled craft. While the USAAF proposal incorporated fewer technical details than the BuAer study, the RAND planners offered some interesting thoughts on the eventual utility of such a craft.

The RAND satellite, for example, would gather scientific data and pave the way for later generations of spacecraft that would revolutionize communications, weather forecasting, and reconnaissance. In addition, as the RAND engineers pointed out, there was "little difference in design and performance" between a rocket capable of launching an Earth satellite and an intercontinental missile. Moreover, the project would have a psychological impact comparable to the "explosion of the atomic bomb." To understand the importance of the program, one had only to "imagine the consternation and admiration that would be felt here if the U.S. were to discover suddenly that some other nation had already put up a successful satellite."[11]

In an effort to force the USAAF to support the navy program, BuAer officials requested an official hearing on the subject before the Aeronautical Board, a War Department panel that encouraged cooperative projects between the services. In view of the high cost of the navy's plan, the technical problems that remained to be overcome, and the refusal of the air force to participate in a joint venture, the board postponed a decision, suggesting instead that the two services continue to fine-tune their individual proposals.

Paper studies continued to evolve over the next year. By 1947–48, however, the air force had lost funding for the satellite effort and had canceled its initial contracts for ICBM studies in order to concentrate on the development of manned bombers and air-breathing cruise missiles. Enthusiasm at BuAer remained high, but with no prospect of a joint program and no hope for funding, navy satellite studies also came to a halt by mid-1948.

This situation was soon to change. In 1949 the desire for improved aerial surveillance of the Soviet Union coupled with the increased potential for delivering lighter weight nuclear weapons by rocket led to another contract with RAND for the study of military satellites. Renewed authorization for the development of the first U.S. ballistic missiles followed in 1950. Still, the opportunity to launch an Earth satellite before 1955 had disappeared, a victim of budget constraints and intraservice rivalry.

In the late 1940s, as army air force and navy planners were envisioning the first generation of spacecraft, the German engineers who had designed, built, and flown the V-2, the most advanced rocket in the world, found themselves transported back to the United States' Old West. Separated from their families, who had stayed behind and were living in a special U.S. government compound at Landshut, Germany, the veterans of Peenemünde settled into quarters not unlike those that had sheltered U.S. cavalry troopers in pursuit of Geronimo seventy-five years earlier. Trying to make the best of it, the Germans cobbled together furniture from scrap lumber and shipping crates and sneaked out of the barracks at night to scrounge scraps of discarded linoleum, which they used to keep sand and grit from blowing up through the floorboards. Having made themselves as comfortable as possible, they then turned their attention to language studies. Some tried to improve their English skills by reading U.S. scientific and technical journals; others listened to the radio for hours at a time, or spent entire Saturdays at the movies in El Paso. One engineer, who had turned to a New Mexican American friend for language instruction, developed a mixed German-Hispanic accent; another, fresh out of flints *(feuerstine)* for his cigarette lighter, walked into a Firestone tire store and asked for a dozen.

Most of the engineers, like Dieter Huzel, "fell in love with the countryside[,] . . . the colorful, vast, open spaces called the Southwest." The early

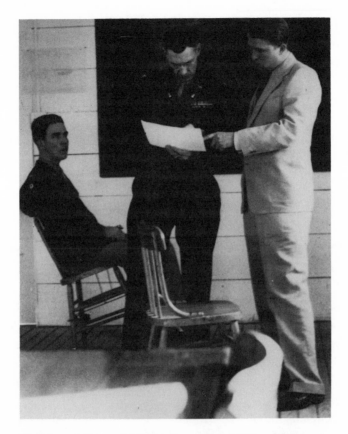

Wernher von Braun *(right)* confers with Col. Holger Toftoy, chief of the U.S. Army Ordnance Branch, at Fort Bliss, Texas, in 1946, as Maj. James B. Hamill looks on. (Courtesy of the National Air and Space Museum, Smithsonian Institution, Photo No. A-4075)

trips from Fort Bliss to the newly established rocket-testing range at White Sands, New Mexico, were made by bus. In the first stage of the journey they traveled north past North Franklin Mountain and up the Rio Grande Valley to Las Cruces, where they might still have found an elderly citizen who had seen Billy the Kid in the flesh. The actual "White Sands," rolling gypsum dunes framed by the stark peaks of the Sacramento Mountains, lay across San Andres Pass in the Tularosa Basin.[12]

When Wernher von Braun and his colleagues first saw the facility, it consisted of one modest hangar and a handful of one-story wooden barracks. Lt. Col. Harold Turner, who had established the test range, had risked his career to achieve even this modest level of development. Unable to find a local source of lumber, he had dismantled a series of abandoned buildings at Fort Bliss and had trucked the materials to White Sands. Only the timely intervention of Col. Holger Toftoy saved Turner from the wrath of a post commander with some very specific plans of his own for the buildings.

Veterans of the most advanced research-and-development facility in the world, the Germans would have to be content with much more primitive working conditions for some time to come. Machine tools and skilled laborers were in short supply; copper tubing and other odd bits of equipment had to be scrounged from a neighboring automobile graveyard. Not long after their arrival at White Sands one of the Germans asked an army officer why only one man in a group of four standing around a machine was working. He was told that the job required metric wrenches, and there was only one available in the entire facility.

In the fall of 1945 the first of three hundred freight cars filled with V-2 parts and pieces that had been gathered at Mittelwerk and elsewhere in Germany began pulling into White Sands. The ordnance officers who had rushed to recover as much as possible from places like Peenemünde before the arrival of the Red Army had worked without the benefit of parts lists or detailed drawings of the rockets: they simply had scooped up everything in sight and hoped for the best.

Reassembling old missiles from the often rusted, corroded, and broken pieces available scarcely seemed worthwhile to the German engineers, who were anxious to move beyond the V-2. But U.S. Army Ordnance Rocket Branch chief Holger Toftoy saw this reassembly work in an entirely different light. The project, he felt, would provide the U.S. military with its first direct experience in handling large ballistic missiles. It also would give ordnance a considerable lead over its rivals in the other services and draw leading industrial and research organizations, such as General Electric, North American Aviation, Douglas Aircraft, the Jet Propulsion Laboratory, or Aerojet General, into partnership with the army.

Toftoy also recognized the potential value of building a relationship between the army's missile program and the nation's scientific community. When Ernst H. Krause of the Naval Research Laboratory (NRL) learned of the army's plan to rebuild and launch V-2s in the desert, he asked Toftoy to consider turning several of the rockets over to the NRL for upper-atmosphere research. Intrigued by the offer, the general urged Krause to contact col-

A captured V-2 arrives at the U.S. Army Ordnance Proving Ground, White Sands, New Mexico, on 10 May 1946. A total of sixty-seven V-2s were launched between 1946 and 1952, giving the United States its first experience with large ballistic missiles. (Courtesy of the U.S. Army)

leagues at other institutions and to create a scientific panel for the selection of instruments and the designation of experiments to be flown on the V-2s.

On 27 February 1946, a date that marks the birth of organized space science in the United States, eleven scientists met at Princeton University to form the V-2 Upper Atmosphere Panel, appointing Krause as its head. Throughout its history, the Upper Atmosphere Rocket Research Panel, as it was renamed in 1947, had only two chairmen: Krause and James Van Allen, a University of Iowa physicist who directed the organization from 1947 until the creation of the National Aeronautics and Space Administration (NASA) in 1958. Up to that time, the research panel enjoyed no official status, no formal charter, and no budget of its own. As Homer Newell, a leading member of the group, explained, "Whatever control [the panel] might bring to bear on the program was exerted purely through the scientific process of open discussion and mutual criticism."[13]

Consisting of representatives of the major universities and laboratories that were engaged in space research, the panel met four or five times a year to evaluate the usefulness of experiments proposed by space scientists, including themselves. It offered suggestions, criticism, and, when it thought the time was right, scheduled experiments for flight. The army, the funding agencies, and the individual scientists themselves accepted the decisions of this unofficial and informal peer-review panel as final.

Colonel Toftoy was delighted with the panel's work. At no cost to his own program, he had been able to forge a firm alliance with the powerful forces of U.S. science. Moreover, the fact that his rockets would be carrying scientific instruments to the edge of space had enormous public-relations value.

Official support for the primary goal—a serious U.S. ballistic-missile program under the control of U.S. Army Ordnance—was limited. Toftoy hoped to generate public enthusiasm for this work by linking the project to the overall importance of science and the excitement of exploring a new frontier. "Mr. Missile," as he became known, never missed an opportunity to remind Americans that the U.S. Army was aiming for the stars. "You are living at the beginning of a new age," he assured young radio listeners, urging them to study hard in school so that they could "be the ones to perfect the rockets of the future—the Buck Rogers kind that go clear to the moon."[14]

A grand total of sixty-four V-2 rockets were sent aloft into the clear, blue New Mexican sky between 16 April 1946 and 19 September 1952. One additional missile was launched from the deck of an aircraft carrier, and two others soared skyward from a new facility at Cape Canaveral, Florida. The first rocket fired from White Sands was a failure, and the V-2's record remained spotty for some time. In May 1947 three of the four V-2s launched went out of control, with potentially disastrous consequences for the citizens of two nations: one landed less than 5 miles (8 kilometers) from Alamogordo, New Mexico, and another within three miles (five kilometers) of downtown Las Cruces; the third created a crater 30 feet (9 meters) deep and 50 feet (15 meters) across when it slammed to earth near a cemetery a mile and a half (2.4 kilometers) south of Ciudad Juarez, Mexico.

The V-2s flown at the outset of the program essentially were wartime originals. Gradually, however, modification became the order of the day. By 1950 a White Sands V-2 stood 5 feet (1.5 meters) taller than its predecessors had in 1946 and featured a much larger payload compartment. On 22 August 1952 a V-2 launched for maximum altitude reached a height of 133 miles (213 kilometers) above the earth's surface.

The real quest for altitude, however, began with Project Bumper, in which a WAC Corporal rocket was mounted as a second stage on the nose of a V-2. The WAC (for Women's Auxiliary Corps) Corporal, so named to identify it as a "little sister" of the Jet Propulsion Laboratory's Corporal E design, was the first U.S. high-altitude research rocket and was capable of boosting a 25-pound (11.4-kilogram) instrument payload to an altitude of more than 100,000 feet (30,500 meters). Conceived by Frank Malina and designed by JPL, the rocket was propelled by an Aerojet General engine. The Douglas Aircraft Company handled its final assembly.

Malina and his colleague Martin Summerfield had suggested the idea of a two-stage V-2 WAC Corporal to U.S. Army Ordnance in 1946. Eight launches were conducted between 13 May 1948 and 29 July 1950. On 24 February 1949 a Project Bumper rocket reached a record altitude of 244 miles (390 kilometers). At long last, human beings had sent one of their own creations into space.

Things were moving far too slowly to suit the Germans, but Project Hermes provided some opportunities to push the limits of rocket technology. The Hermes family of missiles included the A-1, a surface-to-air vehicle similar to the German Wasserfall, developed late in World War II; a solid-propellant A-2; the disappointing A-3A and A-3B; the Hermes B, a supersonic, ramjet-powered second stage for the V-2; and the Hermes C, a "dream rocket" designed to boost an unpowered glider a distance of 2,000 miles (3,200 kilometers). With the exception of the A-1 and A-3, however, none of these projects resulted in flight vehicles. And yet the work was by no means a waste: the A-3 rockets led to improved engines and the development of sophisticated systems for guiding missiles by radio and through inertia, or the use of gyroscopes. Von Braun himself dated the origins of the successful Redstone missile to his work on the Hermes C program.

By the end of the decade the V-2s were being superseded by a new generation of specialized research rockets. In 1946 Ernst Krause and Milton Rosen of the NRL began to develop plans for the Viking. A large research rocket designed to replace the V-2, the Viking was produced by the Glenn L. Martin Company. Reaction Motors, a firm established by a group of American Rocket Society experimenters, won the contract for the 20,500-pound-thrust (91,200-newton) motor. In the end fourteen Vikings were launched between 3 May 1949 and 1 May 1957. They returned a great amount of information and led directly to the first U.S. satellite effort.

Aerobee, another new sounding rocket, grew out of the WAC Corporal program. Developed under the supervision of staff members at the Applied Physics Laboratory, Johns Hopkins University, and the Office of Naval Research (ONR), the rockets were produced by Aerojet General. They were designed to carry 100 pounds (45.5 kilograms) of scientific instruments to an altitude of 75 miles (120 kilometers) and found a market with the army and air force as well as the navy. Modified, improved, and redesigned into an entire family of rockets (Aerobee-Hi and Aerobee 75, 100, 150, 200, 200A, 300, 300A, and 350), the Aerobee ranks as the most successful sounding rocket of the space age. The launch of the final Aerobee (number 1,058) on 17 January 1985 marked the end of an era.

With contracts for its new research rockets and rocket-propelled-takeoff units in hand and the use of rocket weaponry on the rise, Aerojet looked forward to a bright future as it entered the 1950s. Unfortunately, the new decade would be less kind to the little band of Caltech rocketeers who had founded the company in 1942.

From the outset Theodore von Kármán had been overwhelmed by the pressures of trying to meet wartime orders and handle growing credit problems while keeping up with his university duties and research commitments. In the spring of 1944 Aerojet principals instructed Andrew Haley, their attorney, to look for new investors who would be willing to assume some of the burdens of management. The General Tire and Rubber Company of Akron, Ohio, extended $500,000 in credit to the firm and accepted responsibility for the fulfillment of existing navy contracts.

Von Kármán, Malina, Summerfield, and Haley countered by offering to sell 50 percent of their stock in the firm to General Tire for $225,000; General Tire came back with an offer of $50,000. Anxious to escape their management problems and recognizing that their navy customers would feel safer in more experienced hands, the disappointed founders of Aerojet settled for $75,000. As General Tire already had acquired the shares owned by Parsons and Forman, the Akron firm now held controlling interest in Aerojet, which it renamed the Aerojet General Corporation.

But the end of their time as the managers of Aerojet did not mark a return to research and teaching for the GALCIT team. Von Kármán, now an esteemed national figure, was absent in Washington, D.C., assisting USAAF officials plan for a future that would be shaped by technology. The burden of administering GALCIT fell on Frank Malina's shoulders by the end of 1944. His life became a hectic round of trips between California, Wright Field, and Washington. "Life," he later recalled, became a "between trips kind of existence."[15]

Perhaps more important, in the months following the end of World War II Malina found weapons research "more and more distasteful." Convinced that "war between or by states with advanced technology was a form of national insanity," the thirty-four-year-old engineer decided to dedicate his future to finding "ways for 'sovereign' nations to function in peace together, rather than to develop better means of destroying themselves."[16]

He requested and received a two-year leave of absence from JPL in 1947 and traveled to Paris to begin work with the Natural Sciences Section of the United Nations Educational, Scientific and Cultural Organization (UNESCO). Malina would remain involved in world scientific organizations until his death in 1981, in addition to earning a second reputation as a successful painter and sculptor. Malina was universally recognized and honored as a leading space pioneer, but his days as one of the world's most active rocket researchers had come to an end in 1947.

The situation for some leading members of the old "suicide squad" who had remained at Caltech was turning very sour indeed. The problems began in 1949, when a series of startling revelations rocked U.S. society to the core. In September of that year the Soviets shattered the United States' nuclear monopoly by exploding their first atomic bomb. The following month Alger Hiss brought suit to counter assertions that he had been a communist agent. In December Chiang Kai-shek abandoned mainland China to the followers of Mao Tse-tung. Two months later, in February 1950, British officials admitted that British physicist Klaus Fuchs had sold vital nuclear secrets to the Soviets. Later that month Senator Joseph McCarthy announced that he had uncovered 205 Communists in the State Department. On 25 June 1950 forces of the Democratic People's Republic of Korea poured across the thirty-eighth parallel into South Korea.

The repercussions of those events reverberated across the nation. In June 1950 Hsue Shen Tsien, Goddard professor of jet propulsion at Caltech, was informed that his security clearance had been revoked. The FBI believed that a political discussion group to which both he and Malina had belonged in the mid-1930s had actually been a Communist Party cell. When efforts of influential friends, including Caltech president Lee DuBridge, failed to clear his name, the scientist and his family attempted to return to China. Tsien was seized by U.S. authorities at the airport, jailed for a time, and then released. He was finally allowed to continue teaching but was denied access to secret information.

Suddenly, five years after his security clearance had been revoked, Tsien was told that he could leave the United States. He sailed for China with his family on 17 September 1955. Over the next two decades he became one of the most important scientific figures in China. As Qian Xue Sheng, he rose to the position of chief designer of the Chinese space program. As such, he was ultimately the most successful member of the Caltech crew.

John Parsons went out with a bang, as he surely would have wished. He was at work in his mother's carriage house on South Orange Avenue in Pasadena on the afternoon of 17 June 1952, when an enormous explosion rocked the neighborhood. Forensics experts later concluded that he had dropped a container of fulminate of mercury. Neighbors who rushed into the demolished structure noticed a strange smell in the air. Perhaps it was nothing more than a lingering whiff of brimstone. Frank Malina, ever faithful to his friends, saw to it that a crater on the Moon was named in Parsons's honor.

And what of William Bollay, the graduate student whose lecture on rocketry had played such an important role in bringing the Caltech rocketeers together in the first place? He joined the navy in 1941 and was assigned to the BuAer rocket research program at Annapolis, where his colleagues included both Robert Goddard, who was at the end of his career as a rocketeer, and Robert Truax, a recent graduate of the U.S. Naval Academy who was just beginning a lifetime dedicated to reaction propulsion. After the war Bollay accepted a position at North American Aviation working on the design of first generation carrier jets. Instead he was drawn into work on an early missile proposal for the U.S. Army Air Force. Given his background, he quickly emerged as the company's expert on rocketry and a central figure in the development of the reaction engines ultimately intended to power the Navaho cruise missile.

The engines, the experience, and research and development facilities that grew out of the project proved to be far more important than the missile. With the 56,000-pound-thrust V-2 power plant as a starting point, the North American engineers developed, by 1950, a 75,000-pound-thrust engine with pathbreaking features that set the stage for further improvements. Rocketdyne, North American's engine division, would produce engines for a string of missiles and launch vehicles, from Thor through the Saturn 1B to the Space Shuttle.

Today a metal plaque at the mouth of Arroyo Seco marks the spot where the members of the "suicide squad" tested their first rocket motors. NASA's Jet Propulsion Laboratory has grown right up to this site, which was once considered so isolated and barren that it could safely be entrusted to the young men who had attacked Caltech's lawns and laboratories with their corrosive chemicals.

The plaque is a useful reminder to passersby of what an earlier generation of Caltech graduate students accomplished, but the giant laboratory itself is their true memorial. At JPL—the home of Explorer 1, Ranger, Mariner, Viking, Pioneer, Voyager, and a dozen other spacecraft that have opened new doors on the universe—the memory of Theodore von Kármán and his pack of brilliant young engineers remains forever green.

7 • Selling Space Flight

t soon was apparent that the German scientists and engineers had come to the United States to stay. Their original six-month contracts were extended once, then replaced by annual agreements amounting to regular employment. The first of the families arrived at Fort Bliss, Texas, late in 1946. A wartime hospital was transformed into somewhat spartan family quarters. Wernher von Braun visited Germany in the spring of 1947, returning to Texas a few weeks later with his parents and a bride, Maria von Quistorp, an eighteen-year-old cousin. Soon thereafter von Braun became the proud owner of a brand-new Nash Ambassador automobile. It was a sure sign he was putting down roots.

Naturalization posed some special problems for these German engineers, who had come to the United States outside of the normal immigration channels, and without visas. In the spring of 1948 they boarded a street car in El Paso, climbed off at the U.S. Consulate in Cuidad Juarez, Mexico, and reemigrated, visas now in hand, a few hours later.

Just two years later Col. Holger Toftoy flew von Braun to Huntsville, Alabama, the new home of the U.S. Army Ordnance rocket program. Nestled in a bend of the Tennessee River, Huntsville was a quiet place whose sixteen thousand citizens took considerable pride in their city's twin claims to fame as watercress capital of the world and the hometown of actress Tallulah Bankhead.

Huntsville had enjoyed a century and a half of slow growth as a cotton-trading center and mill town before U.S. Army Ordnance and the Chemical Corps established the Redstone and Huntsville arsenals in 1941. The workers at the Redstone plant spent the war years filling a variety of projectiles, from artillery shells to hand grenades, with poison gases and toxic chemicals produced at the Huntsville facility. When the two arsenals closed in 1945 boom times gave way to unemployment and economic hardship. Alabama senator John Sparkman tried and failed to convince the U.S. Air Force to establish a new research facility in Huntsville, his hometown. Toftoy and his rocketeers were to be the consolation prize.

After four years in the desert the Germans promptly fell in love with Huntsville. Although the reciprocal process took a bit longer, Huntsville also fell in love with them. The newcomers were model citizens who, it was said, took out library cards before obtaining their Alabama driver's licenses. Within a few weeks of their arrival they founded a Lutheran church, as well as a Boy Scout troop that produced a record number of Eagle Scouts during its first year of existence.

The Germans also took the lead in local charity drives, and they worked to establish both a symphony orchestra and an amateur observatory. Two years after coming to town, Walter Weisman, a Luftwaffe veteran, was elected president of the Junior Chamber of Commerce. The following year, the daughter of one of the German scientists won the Good Citizenship Award offered by the Daughters of the American Revolution. But their proudest moment came on 15 April 1955, when forty Germans were sworn in as citizens of the United States. The long process that had begun with the decision to abandon Peenemünde and seek a new communal home was complete.

The Germans managed to blend into their new surroundings without sacrificing their own sense of community. Most of them did not have the wherewithal to buy individual homes in Huntsville; they tackled this problem with their usual sense of unity and organization. Fifteen families scraped together the purchase price for a tract of wooded land near the crest of Monte Sano, overlooking the city. Learning that local banks would make a home loan to any individual with a good job and twenty-five hundred dollars in the bank, four families pooled their resources, and, as each engineer in turn applied for his loan, proceeded to transfer the requisite amount from one account to another. They were fairly sure that bank officials understood the scheme and chose to trust them anyway.

The same sense of group cohesion characterized their professional lives.

They had no intention of allowing anyone to milk them of their knowledge and then discard them, or to break up the team that had been so successful. At White Sands they had contributed to the U.S. effort while also insisting that the U.S. Army and General Electric respect the integrity of the group. Some Americans chafed at the formality and rigid chain of command that marked the German rocket team. The efforts of army commanders to fully integrate the team with U.S. contractors failed, and von Braun repeatedly threatened to resign when he felt that his leadership of the Peenemünde group was being questioned.

The U.S. Army rocket program grew around this core of Germans, with von Braun, the unquestioned technical leader, supervising the work of thousands of employees and contractors. The influx of American newcomers did not dilute the German team spirit, however. The old comrades rose with von Braun to become unit heads and division chiefs. In fact, the heart of the team was to remain intact for a quarter of a century after the end of World War II.

In the summer of 1950, not long after their arrival in Huntsville, the army charged von Braun's team with a single task: to develop a ballistic missile capable of delivering a nuclear payload to targets at least 500 miles (800 kilometers) away. The designation of the project and the precise requirements for the missile continued to evolve over the next two years. By 1952 the weapon had become known officially as Redstone, the weight of its payload had been increased, and its expected range reduced to 200 miles (320 kilometers). North American Aviation would produce the engines, the Chrysler Corporation would help manufacture the missile, and the Ford Instrument Company would be involved in developing the all-important inertial-guidance system.

Everyone at the Redstone Arsenal recognized the importance of this project, but the technical challenges involved in taking a relatively small step beyond the V-2 were neither especially demanding nor engaging. Furthermore, the members of the team chafed at the small budget that had been allocated for the project, and at what they regarded as the even more limited—and limiting—vision of U.S. military leaders.

Gradually, a few key German figures were lured away to industry. Krafft Ehricke, Hans Friedrich, and Walter Schwidetsky, for example, accepted jobs with the Consolidated Vultee Corporation (later General Dynamics), which had won the contract to develop the Atlas intercontinental ballistic missile for the U.S. Air Force. Ehricke eventually would become vice president of General Dynamics, with responsibility for advanced projects. Colleagues Walter Dornberger, Adolf Thiel, and Martin Shilling followed similar patterns, becoming the vice presidents, respectively, of Bell Aircraft, TRW, and the Raytheon Corporation. Following his retirement from NASA in 1972 Wernher von Braun accepted a position as vice-president for engineering and development with Fairchild Industries in Germantown, Maryland.

But during the doldrums of the late 1940s and early 1950s, as he had in the past, von Braun kept his workers busy and productive by focusing their attention on the ultimate goal of space flight. He first had dreamed of a flight to Mars at Kummersdorf in the 1930s. Now, in 1947, he began to commit that thinking to paper. *Das Marsprojeckt,* completed the following year, was a bad science fiction novel with a marvelous technical appendix that laid out the exact requirements for such a flight. Failing to interest a U.S. publisher in his manuscript, von Braun finally allowed a German firm to issue the appendix as a single, thin volume in 1950.

There were some signs that U.S. interest in space flight, leavened with a bit of Cold War apprehension, was on the rise. The flying-saucer craze, which began in 1947, spawned a new generation of science fiction films and had everyone looking to the skies and wondering what might be out there. According to a 1951 *Life* magazine estimate there were as many as two million hard-core science fiction fans in the United States. Indeed, the very nature of the genre was changing. The garish covers of the Gernsback pulps had vanished from the newsstands, replaced by magazines featuring stories that explored the social, political, and cultural implications of science in general and of space travel in particular.

Eyeing this expanding market, some of the major publishing houses began to enter the field, and young readers soon were devouring engrossing and well-written science fiction novels by a new generation of authors, including Isaac Asimov, Robert Heinlein, and Arthur C. Clarke. Clarke, Willy Ley, and others also produced nonfiction books on space flight. The magazine and book illustrations of space-flight artist Chesley Bonestell and films such as *Destination Moon* underscored the excitement.

On 12 October 1951 the First Annual Symposium on Space Travel, held at the Hayden Planetarium in New York, ushered in a new era in the promotion of space flight. Organized by planetarium director Robert Coles and writer

Willy Ley, the symposium featured presentations by luminaries such as Fred Whipple of the Harvard Observatory and Heinz Haber of the U.S. Air Force School of Aviation Medicine.

Two writers for *Collier's* magazine attended an evening reception and dinner at the planetarium and returned to Gordon Manning, their managing editor, brimming with enthusiasm for an article on space flight. Intrigued, Manning dispatched Cornelius Ryan to San Antonio for a space-medicine symposium the following month. An associate editor of *Collier's* and one of its most popular writers, Ryan went to the meeting a space-science skeptic and returned to New York a convert. Von Braun, Whipple, and their cohort from the National Academy of Sciences, Joseph Kaplan, had convinced him to write a popular magazine article on space flight. *Collier's* space-flight project expanded to eight feature articles, which appeared between 22 March 1952 and 30 April 1954. Manning and Ryan enlisted the aid of many of the field's leading experts, but von Braun was the project's central figure, writing some of the articles himself and shaping the series as a whole. As the series' potential for spectacular illustrations had been one of its strongest selling points, *Collier's* art director, William Chessman, commissioned the best artists he could find: Chesley Bonestell, Fred Freeman, and Rolf Klep. The series was an overwhelming success, attracting some twelve to fifteen million readers and inspiring *Time, Look, This Week,* and other magazines to feature their own space-flight stories, many of which focused on Wernher von Braun as the person in charge of the U.S. space effort. Viking Press, which had published Willy Ley's *Conquest of Space* (1949) complete with classic illustrations by Chesley Bonestell, transformed the *Collier's* articles into three best-selling books: *Across the Space Frontier* (1952), *Conquest of the Moon* (1953), and *The Exploration of Mars* (1956). Von Braun produced other space-flight books on his own, including one, *A Flight to the Moon* with illustrations by Fred Freeman, that was aimed at younger readers.

In 1954, as *Collier's* ran the last of the space-flight articles, animator Walt Disney was drawing up plans for a "magic kingdom" that would rise from the orange groves near Anaheim, California. The need to publicize this venture led him into another new technological field: television. Disney had signed a contract with the American Broadcasting Corporation (ABC) network for a series of "Disneyland" programs to be aired on Sunday evenings beginning in September 1954. The programs would represent the new park's three theme areas: Fantasyland, Frontierland, and Tomorrowland. Longtime Disney animator and associate Ward Kimball, who was in charge of developing Tomorrowland ideas for both the park and television, had taken note of the illustrations in the *Collier's* series. He suggested to Disney that they contact Wernher von Braun for advice and assistance. Once again, the German engineer wound up as the star of the show.

In the end three space-flight programs, each based on ideas developed by the rocket team at Huntsville, appeared on the "Disneyland" show: "Man in Space," "Man and the Moon," and "Mars and Beyond." Recognizing von Braun's innate charisma and appeal, Disney producers employed him throughout the series as the leading on-screen personality. The programs had an enormous impact. The Disney animators and the handsome young rocket engineer with the wavy hair and broad shoulders made the once-outrageous business of space flight seem very real indeed. Millions of young Americans would remember that the excitement of space flight had been revealed to them in a soft, rolling German accent.

By 1955 Wernher von Braun had emerged as one of the most recognizable figures in U.S. science and technology, his face and voice familiar to millions. For all the adulation he received, however, the public's perception of him remained somewhat confused. If most Americans admired actor Curt Jurgen's portrayal of von Braun in the 1960 Columbia Pictures release *I Aim at the Stars,* others saw him as the model for the unregenerate Nazi turned U.S. military advisor, *Dr. Strangelove.* Producer Stanley Kubrick once asked Arthur C. Clarke to assure von Braun that Kubrick "wasn't getting at him" with the Strangelove character. "I never did," Clarke explained, "because (a) I did not believe it (b) even if Stanley wasn't, Peter Sellers most certainly was."[1]

Von Braun's achievements were recounted in a number of inspirational biographies aimed at young people, and in the ironic lyrics of a Tom Lehrer song: "If the rocket goes up, who cares where it comes down / 'That's not my department,' says Wernher von Braun."[2]

They did not sing songs about Sergei Pavlovich Korolev. He was a legend in the Soviet Union, a figure of heroic stature who had led his nation to the very peak of postwar power, glory, and prestige. Yet his face did not appear on television or on the covers of the national news magazines, and no one outside his small circle of friends knew his name. The powers that be in the Kremlin insisted that he remain an anonymous figure: the chief designer.

Korolev and von Braun were never to meet, although they once came

Walt Disney *(left)* **with Wernher von Braun, who emerged as a media star in 1955 when Disney invited him to assist in the development of three "Disneyland" television shows about space flight. These shows created new interest in space among America's postwar generation. (Courtesy of the National Air and Space Museum, Smithsonian Institution, Photo No. 80-19168)**

close. A number of Soviet officers, including former Gas Dynamics Laboratory rocketeer Col. Valentin Petrovich Glushko, had applied for and received permission to witness the Allied Operation Backfire V-2 firings at Cuxhaven, Germany, which were conducted from August to October 1945. The official Soviet observers arrived on schedule, along with a number of additional personnel who had not received clearance. Despite strong Soviet protests the "unofficial" members of the party were not allowed to enter the launch complex and had to watch the proceedings from outside the fence. One of these beyond-the-fence watchers was Sergei Korolev.

Postwar Soviet strategists would, at long last, come to share Korolev's enthusiasm for rockets. Enmeshed in the early stages of a Cold War with the West, the USSR was surrounded by an arc of potential enemies that stretched from Scandinavia to Turkey and was within possible striking range of a new generation of giant U.S. bombers—the Convair B-36. Their principal targets, on the other hand, were half a world away in the United States. Soviet scientists were making great progress with a nuclear-weapons program, but there was no practical delivery system for these weapons at hand. The V-2 was impressive, Marshall Zhigarev, chief of the Red Air Force commented in 1946, "but what we really need are long-range, reliable rockets capable of hitting the American continent." Engineer A. G. Kostikov summed the matter up: "Everyone wanted to design a trans-Atlantic rocket."[3]

In 1947 G. A. Tokaty-Tokaev, an air force researcher and member of a special state commission on long-range rocketry, was ordered to brief members of the Politburo on Austrian Eugen Sanger's wartime design study for an "antipodal bomber." Nicknamed "the Great Silver Bird," it was a concept for a rocket-propelled lifting body that would climb to an altitude of 160 miles then travel transcontinental distances by skipping across the top of the atmosphere. Tokaty-Tokaev was asked if he thought the Americans would be interested in so wild a dream. "If it be true that the Americans are so concerned with rocket weapons that they have transformed Texas into a vast Peenemünde," he replied, "it is hardly possible that they have overlooked Sanger's plan."[4]

Although the Soviets, like the Americans, would have to walk before they could run, German brains and hardware would provide them too with a postwar jump-start. Helmut Gröttrup, who had served as an assistant to the chief of the guidance-and-control section at Peenemünde, was the highest-ranking member of the von Braun team to join the Soviet program. With one Peenemünde veteran in hand, the Soviets had scoured East Germany for lower-level members of the German rocket program. Gröttrup's first assignment was to assist Glushko, Korolev, and other Soviet experts in putting the V-2 fabrication plant at Nordhausen, and a handful of other wartime rocket-part factories, back into production. Components that could not be manufactured in the Soviet bloc would be obtained by hook or crook from suppliers in Western Europe.

The scheme worked. In September 1945 the Soviets began static-testing V-2 motors. By the fall of 1946 they had assembled some thirty missiles for testing. It soon became clear, however, that funding a rocket-research institute in East Germany and setting the members of the "Gröttrup team" to work producing obsolete V-2s were not the most effective ways to build Soviet capability in the field. On 22 October 1946 the German employees of the Soviet rocket program learned that they were being forcibly removed to the Soviet Union. Once there, they were split up, sent to different locations, and put to work on separate development projects, training Soviet colleagues in the process.

The Germans remained on good terms with their Soviet superiors, but there was never any chance that these Peenemünde veterans would become the master technicians of the Russian rocket program. The first significant product of postwar Soviet rocketry was a missile was the R-1, or Pobeda ("victory"). About the same size as a V-2, it nevertheless represented a considerable advance over that missile. The German original had a range of 190 miles (300 kilometers), an empty weight of 8,836 pounds (4,016 kilograms), and tipped the scales at just over 28,000 pounds (12,700 kilograms) fully fueled and ready for launch. The R-1 could reach targets as far as 570 miles (900 kilometers) away, had an empty weight of 4,225 pounds (1,920 kilograms), and had a launch weight of almost 41,000 pounds (18,600 kilograms). The secret of the rocket's success was in the use of a warhead that separated from it in flight, as well as its monocoque structure. The missile's skin doubled as propellant tanks, a weight-saving technique that later would be applied by Convair engineers in the design of the U.S. Air Force's Atlas missile. The R-1's motor, the RD-100, was based on that of the V-2 but developed almost 14,400 additional pounds of thrust (64,057 newtons).

Test launches of the R-1 began on 18 October 1947 at the original Soviet test site at Kapustin Yar, 60 miles (96 kilometers) east of Stalingrad (now Volgograd). Before the end of 1950 Pobeda missiles were in the hands of the first operational unit of the Rocket Troops, a predecessor of the Strategic Rocket Forces, the Soviet Ministry of Defense organization responsible for all rocket launches, military and scientific, from 1960 to 1991.

By 1949 the Soviets were working on a variety of rockets and assigning the German team to a range of tasks, including the development of a family of research rockets based on the V-2 and general design studies for advanced rocket weapons. The Germans devoted much of their effort to the design of an R-14 rocket capable of boosting a 6,600-pound (3,000-kilogram) warhead to a maximum range of 1,800 miles (2,880 kilometers), but the vehicle was never constructed.

The German experience influenced the early Soviet rocket program, just as it did the U.S. effort. Basic motor-design philosophy, the use of gases from the combustion chamber to operate the turbopumps, small vernier rockets for roll control that were fed by exhaust gases from the turbopumps, and swiveling combustion chambers were all ideas that had first been considered by the advanced-project group at Peenemünde. Now these ideas were being employed on both sides of the Iron Curtain.

In the late 1940s, however, the Soviet rocket effort seems to have passed through a crisis. Living conditions deteriorated for the Germans during their stay in the Soviet Union. Alcoholism was a major problem. There were broken marriages and an alarming number of suicides. The team spirit that sustained von Braun's colleagues in the United States collapsed in the USSR. The Soviets removed Gröttrup from his position of authority. Some German engineers broke away and attempted to find places for themselves on regular Russian rocket-development teams. The end came on 15 November 1953, when most of the German rocket-team members were informed that they were going home to East Germany. The official announcement from Soviet authorities closed with a simple expression of gratitude: "We take this opportunity to express our thanks for the work done."

The Soviet engineers faced their own set of bureaucratic problems. In the best Stalinist tradition, the rocket program was plagued by bitter rivalries. Prior to the Stalinist purges of the 1930s Soviet rocketeers had kept pace with the German army program. The years of terror and wartime hardship had destroyed the prewar program but not dimmed the enthusiasm of the small band of engineers who continued to dream of space flight. Two Ukrainian survivors of the golden age of GIRD, GDL, and NRII emerged from the chaos of war with their enthusiasm, and rivalry, intact. The first, Sergei Korolev, was the best known of the prewar rocket builders/team leaders. A larger-than-life figure, he was an inspirational leader and a cunning politician. The second, the brilliant Valentin Glushko, had built his reputation as a designer of rocket engines and an adept manager.

The old rivals were joined by two emerging figures, also Ukrainians. Vladimir Chelomei was a brilliant theoretician and analyst and a past master at bureaucratic maneuvering. Volunteering to develop a Soviet version of the

V-1 "buzz bomb" in 1944–45, he would spend the next twenty years developing cruise missiles that would rival Korolev's rockets as a potential delivery system for nuclear weapons. Mikhail Yangel, an aircraft design and production specialist, and a dedicated party member, was to replace Glushko as Korolev's principal postwar rival and sometime supervisor. The situation began to ease in 1954, when the Soviet leadership, concerned that military rockets were being produced by only one vulnerable design bureau, placed Yangel in charge of a second bureau specializing in ballistic missiles utilizing storable propellants. Yangel's widow, interviewed by James Harford, Korolev's American biographer, explained that "Korolev work[ed] for TASS [the Soviet news agency] and Yangel for us [the Soviet military]."[5]

The Soviet rocket program moved forward despite political hazards and interpersonal feuds. By 1952 work was underway on the R-2, the first Soviet intermediate-range ballistic missile (IRBM), with a range in excess of five hundred miles (eight hundred kilometers). The subsequent R-2A version stood almost fifty-eight feet (eighteen meters) tall and was powered by another V-2-type engine, the RD-101, which delivered 80,000 pounds of thrust (355,872 newtons). In 1953 a new state organization, the Ministry for Medium-Machine Building, was established to plan and direct the manufacture of rockets and missiles.

Sergei Korolev profited from then–Soviet leader Nikita Khrushchev's early predisposition to favor victims of the late Stalinist purges. Korolev joined the Communist Party and was elected to the Soviet Academy of Sciences, an organization of only 150 members. "We had absolute confidence in Comrade Korolev," Khrushchev recalled in his memoirs. "When he expounded his ideas, you could see passion burning in his eyes, and his reports were always models of clarity. He had unlimited energy and determination, and he was a brilliant organizer."[6]

Korolev had reemerged at precisely the right moment. In view of the difficulties encountered with both postwar bomber and winged-missile programs, Soviet strategic planners now saw the ICBM as the most promising means of delivering nuclear weapons on distant U.S. targets. Korolev found that the members of the Politburo were easy converts to his cause. "I don't want to exaggerate," Khrushchev remarked, "but I'd say we gawked at what he showed us as if we were sheep seeing a new gate for the first time." Khrushchev admitted that on their first visit to a launch pad he and his colleagues had behaved "like peasants in a marketplace. We walked around the rocket, touching it, tapping it to see if it was sturdy enough—we did everything but lick it to see how it tasted."[7]

Indeed, Korolev and his colleagues had entered a new golden age. In 1955, on the vast and empty steppes of Kazakhstan in Central Asia, they inaugurated a new launch complex near the little town of Tyuratam. The architectural remnants of a culture that had been celebrated in the popular Broadway musical *Kismet* were swept away and replaced by giant launch stands, blockhouses, research laboratories, assembly buildings, housing and shopping complexes, and a university. Soviet bureaucrats of the sort who had named the rocket building organization Ministry for Medium-Machine Building insisted that the new facility be officially known as the Baikonur Cosmodrome, despite the fact that the town of Baikonur lay many miles distant. Western analysts were amused but not deceived.

But the rockets flown from Tyuratam and from the original site at Kapustin Yar, and the new generation of motors that propelled them, were no laughing matter. Western intelligence analysts paid close attention to the early short- and medium-range ballistic missiles they called the SS-1 Scud and SS-3 Shyster. The Soviets' next step was clear. By 1955 their rocketeers, with the enthusiastic support of the government, were at work on their first-generation ICBM/space-launch vehicle.

Soviet rocket designers approached this task with a handicap: their nuclear weapons were larger and heavier than those of their U.S. rivals. It would take enormous thrust to lob such hefty warheads half way around the globe. With no single engine on hand capable of meeting the challenge, Russian engineers had little choice but to cluster a series of less-powerful motors.

Glushko and Alexei Isayev spearheaded the development of two such rocket motors. Apparently designed in 1952, the RD-107 broke from the V-2 design tradition. A single turbopump fed liquid oxygen and kerosene to four clustered combustion chambers producing almost 225,000 pounds of thrust (1,001,000 newtons). Twin gimbal-mounted vernier motors served as steering rockets. The RD-108, the second motor, was of similar design but produced some 211,680 pounds of thrust (941,637 newtons) for a longer period of time and featured four verniers positioned around the four main combustion chambers.

By 1955 the members of Korolev's team had designed a new launch vehicle around the new motors. North Atlantic Treaty Organization (NATO) intelligence services referred to it as the SS-6 Sapwood. The Soviets officially

Russian technicians inspect the engines on the R-7, or SS-6, the world's first ICBM/launch vehicle. Known to the Russian launch crews as Semyorka, or "old number seven," the R-7 consisted of a single RD-108 core surrounded by four RD-107 clusters, each with its own faring so it could be discarded when its fuel was spent. (Courtesy of the National Air and Space Museum, Smithsonian Institution, Photo No. 80-17297)

identified the new rocket as the R-7. The rocketeers themselves simply called it Semyorka, the Russian equivalent of "old number seven." Whatever one called the missile, it was the world's first ICBM. The core featured a single RD-108 surrounded by four RD-107 clusters, each in its own fairing so that it could be discarded when the propellants were exhausted. With all thirty-two thrust chambers blazing at the moment of launch, the giant rocket, almost 100 feet (30.5 meters) tall, would be boosted away from the pad by 1.1 million pounds of thrust (5,003,238 newtons).

If and when it flew, Semyorka would dwarf Wernher von Braun's Redstone, which was first launched from Cape Canaveral, Florida, on 20 August 1953. Redstone's single motor developed only 75,000 pounds of thrust (333,630 newtons) for 121 seconds. Unaware of Soviet plans for a super-booster, the von Braun team at the Guided Missile Development Division of the Ordnance Missile Laboratories, Redstone Arsenal, saw great possibilities in the Redstone. The missile represented the Huntsville group's first step beyond the V-2 and, at long last, placed the United States within striking distance of space flight. With some modifications, including the addition of existing solid-propellant upper stages, Redstone could be transformed into a satellite-launch vehicle.

In the months immediately following the end of World War II, American military leaders had rejected the long-term potential of the ICBM and the Earth-orbiting satellite in favor of more traditional technologies, such as the jet bomber and the winged cruise missile. By 1953, however, this situation had changed radically. The Soviets now had nuclear weapons in hand, and they seemed to be rushing toward the development of an ICBM.

In September 1954 President Dwight D. Eisenhower appointed forty-six leading U.S. scientists, engineers, and strategists to a Technological Capabilities Panel and charged them with suggesting ways to prevent a surprise attack by the USSR. Chaired by MIT president James R. Killian, the panel pointed to the need for intercontinental and intermediate-range ballistic missiles to counter the potential Soviet threat. The government lost little time in implementing the group's recommendations. Late in 1955 von Braun's team was ordered to begin work on the Jupiter program, aimed at the development of the first U.S. IRBM. Shortly thereafter the air force awarded the Convair division of General Dynamics a contract for the design and production of the Atlas ICBM.

The most deeply classified section of the Killian report was prepared by a subcommittee headed by U.S. inventor Edwin D. Land. Insatiably curious and blessed with a boundless imagination, Land had become a legend in business and science as the man who had invented and marketed the Polaroid technique of instant photography. Having made his fortune, he proceeded to acquire 164 patents for inventions that ranged from 3-D movies to infrared searchlights and guided-missile components. Killian saw Land as the natural leader of the effort to devise ways to bolster the nation's intelligence capability. The recommendations offered by Land's team remain classified, but there is every reason to suspect that they called for the creation of a reconnaissance-satellite system.

Indeed, reconnaissance was the overriding concern of postwar strategic thinkers. Beginning in the late 1940s, camera-laden balloons were sent floating across the roof of the Soviet Union, and U.S. and British aircraft repeatedly brushed against—and flew over—Soviet borders. These efforts provided some useful information but also sparked strong protests from the Soviets and resulted in the loss of U.S. aircraft and crews. By the late 1950s high-flying Lockheed U-2 spy planes were penetrating Soviet air space, but as the capture in 1960 of U.S. pilot Francis Gary Powers demonstrated, reconnaissance aircraft were a less-than-satisfactory solution to the problem.

As early as 1949 Rand Corporation strategists had identified the reconnaissance satellite as the ideal intelligence tool, although they recognized that such a device might violate international law. More important, the presence of an all-seeing eye relentlessly passing overhead might enrage the Soviets and provoke a military response. But a purely scientific satellite with no cameras on board and with an orbit that did not overfly the USSR might prepare the way, both technologically and diplomatically, for the coming of a true reconnaissance spacecraft. By 1954 the perfect excuse for such a program was at hand.

Since 1950 a number of scientists in England and the United States had proposed scientific satellites as a means of extending sounding-rocket studies of the upper atmosphere. One of the most interesting of these proposals had come from University of Maryland physicist S. Fred Singer. In 1953 Singer unveiled his MOUSE (minimum orbital unmanned satellite of Earth). Complete with instruments, batteries, and telemetry equipment, the little craft would weigh no more than 100 pounds (45.5 kilograms).

One of the first serious steps toward a U.S. satellite effort was taken on 23 June 1954, when Frederick C. Durant III invited Wernher von Braun and some other well-placed space enthusiasts to a meeting in Washington, D.C. A wartime naval aviator, Durant had emerged as a leader among American postwar engineers who took the dream of flight beyond the atmosphere very seriously indeed. Durant had served as president of both the American Rocket Society and the new International Astronautics Federation, and had been employed as a rocket expert by Bell Aircraft, the Arthur D. Little Corporation, and the Central Intelligence Agency (CIA). For some time he had been discussing the possibility of a scientific satellite with Cmdr. George W. Hoover of the Air Branch of the Office of Naval Research. The two of them had concluded that the time had come for action.

The Project Orbiter team pulled together by Durant and Hoover included von Braun, Singer, Harvard astronomer Fred Whipple, Aerojet General's David Young, and Alexander Satin, chief engineer of the Air Branch at ONR. The group determined that an army Redstone rocket equipped with small,

Members of the Project Orbiter Committee meeting in Washington, D.C., on 17 March 1955. *Seated, clockwise from left:* Cmdr. George W. Hoover, Office of Naval Research (ONR); Frederick C. Durant III, Arthur D. Little, Inc.; James Kendrick, Astrophysics Development Corp.; William A. Giardini, Alabama Tool and Die; Phillipe W. Newton, Department of Defense; Rudolf H. Schlidt, Army Ballistic Missile Agency (ABMA); Gerhard Heller, ABMA; and Wernher von Braun, ABMA. *Standing, from left:* Lt. Cmdr. William Dowdell, U.S. Navy; Alexander Satin, ONR; Cmdr. Robert Truax, U.S. Navy; Liston Tatem, IBM; Austin Stanton, Varo, Inc.; Fred L. Whipple, Harvard/Smithsonian Astrophysical Observatory; George Petri, IBM; Lowell Anderson, ONR; and Milton Rosen, Naval Research Laboratory. (Courtesy of the National Air and Space Museum, Smithsonian Institution, Photo No. A-2146)

solid-propellant upper stages could orbit an ONR satellite weighing five to seven pounds (2.3 kilograms to 3.2 kilograms). Impressed, navy officials approved preliminary discussions on a joint project with army general Holger Toftoy at Redstone. Handicapped by a tiny budget, however, the Project Orbiter proposal limped forward, reaching the desk of the assistant secretary of defense for research and development in January 1955. However promising, Project Orbiter was not to be.

Scientists around the world were planning an International Geophysical Year (IGY), 1957–58, during which special efforts would be made to study the polar regions, the depths of the ocean, and the upper reaches of the atmosphere. In the United States IGY planners, who hoped to launch a satellite as part of their program, were not impressed by the small size of Project Orbiter's payload. In addition, the military services now were offering rival proposals. The air force proposed launching a relatively large satellite with its as-yet-untested Atlas missile. Milton Rosen and his colleagues at the Naval Research Laboratory countered with a design for a new three-stage satellite launcher called Vanguard.

Growing military and scientific support for a satellite program, the Killian report, and the judgment of CIA officials that the Soviets would soon have the capacity to orbit their own satellite, convinced President Eisenhower to establish such an effort as a national goal during the IGY. Assistant Secretary of Defense Donald Quarles, charged with selecting one of the existing proposals, named a committee headed by Homer Stewart of the Jet Propulsion Laboratory to make a recommendation.

The committee quickly rejected the USAF proposal, which was based on a large and complex launch vehicle that had yet to fly. Although Vanguard would also be an untried booster, it was based on existing technology. The first stage would be similar to the NRL/Martin Viking sounding rocket, and the second-stage motor would be based on that of the Aerobee-Hi. The modified army Redstone booster that von Braun proposed for Project Orbiter was the safest bet, but the payload was very small. Stewart himself voted for Orbiter, but the majority of the committee chose Vanguard.

Over the next three years U.S. citizens and the rest of the world would hear a great deal about Project Vanguard. President Eisenhower insisted that, in spite of NRL involvement, the program was a civilian effort with purely scientific objectives. Every phase of the operation was conducted in full view of the public. The United States, postwar standard-bearer of freedom, would lead the way into a new age, with the rest of the world applauding from ringside seats. That, at least, was the plan.

8 • Fellow Travelers

Sometimes it's better to be a fast second. *Herman Kahn*

By the fall of 1956 Wernher von Braun had emerged as the most visible spokesman for a strong national space effort. He stood at the head of the largest and most experienced team of rocket builders in the world. Each morning, more than two thousand men and women, 80 percent of the total work force of the Army Ballistic Missile Agency (ABMA), reported for duty with von Braun's Guided Missile Development Division. The Redstone rocket was ready for service, and work was underway on the Jupiter, an intermediate-range ballistic missile intended for duty with both U.S. Army units and surface ships of the U.S. Navy.

At the same time, von Braun could feel the technical leadership of U.S. rocketry slipping through his fingers. Having lost his bid for control of the U.S. satellite effort, he suddenly found himself caught in a web of intraservice competition that threatened to destroy his team and to drive the ABMA out of the long-range rocket business.

The rival missile programs had popped up very quickly. After a brief flurry of early postwar interest the U.S. Air Force had dismissed work on large rockets in favor of manned bombers and cruise missiles, such as the Matador, Snark, and, especially, North American Aviation's Navaho. The disappointing performance of these winged missiles, however, as well as the development of lighter weight nuclear and thermonuclear weapons and the acquisition of intelligence reports regarding rapid Soviet progress with large rockets, led to a reassessment of the situation.

In February 1954 Harvard mathematician John von Neumann was asked to head a special Air Force Strategic Missile Panel. The group recommended the immediate development of an ICBM to counter the Soviet threat. The USAF promptly established the Western Development Division (WDD), which was attached to its Air Research and Development Command and staffed with a team of officers and engineers whose job it would be to bring the USAF into the missile age as rapidly as possible.

The result of the WDD's work was a three-engine, liquid-propelled, stage-and-a-half rocket, the Atlas. Developed by Convair, under the leadership of engineer Karel Bossart, it was a radically innovative design, with a very thin skin and propellant tanks that were integral to the airframe. The finished rocket featured a structure so light that, standing unfueled and in the upright position, it would have to be pressurized to prevent it from collapsing under its own weight. This unlikely design, powered by rocket engines that grew out of the Navaho effort, proved to be highly successful. Atlas variants are still in use as workhorse satellite boosters in the 1990s.

The procedures developed to manage Atlas's design and production management were as radical as the missile itself. Under the leadership of Gen. Bernard Schriever, the WDD employed a systems-engineering approach that enabled USAF planners to coordinate the efforts of a host of contractors and subcontractors across the nation who designed and manufactured the rocket's many complex elements. The simultaneous development of myriad high-technology bits and pieces that would have to fit and function together was an extraordinary challenge. Meticulous planning, close supervision, and an emphasis on quality control streamlined the process and reduced the need for extensive testing of the finished product.

The Atlas program moved forward at a rapid pace, but the costs were very high, in part because of the need to provide redundant systems and to pursue alternative approaches that could be employed in the event a primary system proved unsatisfactory during the development and testing phase. By May 1955 many of the concepts that had been worked out but not required for the Atlas were being incorporated into the design of an even larger ICBM, the Titan. In addition, Schriever's command was charged with creating the Thor

The U.S. Air Force's Atlas rocket had a structure so light it would collapse under its own weight unless pressurized. But the design proved so successful that, forty years after entering service, variants continue to boost satellites into orbit. (Courtesy of NASA)

IRBM and was given responsibility for all USAF reconnaissance-satellite and space projects.

In the fall of 1956, understandably nervous at the prospect of operating large, liquid-propellant rockets aboard their vessels, U.S. Navy officials announced their desire to withdraw from the Jupiter program and create a solid-propellant IRBM of their own. The navy followed a path similar to that of the air force: a special-projects office under the command of Rear Adm. William "Red" Raborn would apply systems-management procedures to coordinate and guide the contractors, subcontractors, and navy researchers involved in producing a fully integrated weapons system that would include both the solid propellant Polaris missile and the first generation of "boomers," the nuclear-powered missile submarines that would carry the rockets to a launching point near an enemy's coast.

Compared to the streamlined management style that had been adopted by the other services, the Army Ballistic Missile Agency's approach to research and development was precisely what it appeared to be: an artifact of the nineteenth century. In the earliest days of the Republic the army had produced its own muskets and side arms in government-owned-and-operated factories. As late as World War II specialized weapons still were manufactured in arsenals like Redstone in Alabama.

Wernher von Braun felt most familiar and comfortable with the arsenal system. Redstone Arsenal was a natural product of the conjunction of army tradition and the availability of an entire team of experienced rocketeers. At Huntsville, as at Peenemünde, design, development, and testing were the work of in-house experts, and contractors were responsible for the actual manufacturing of the operational missiles.

Ironically, despite his charismatic leadership, breadth of vision, and reputation as a spokesperson for the future, von Braun was a cautious manager and a very conservative engineer. He had the utmost confidence in the judgment of his division chiefs, many of whom had been with him since Peenemünde, and he insisted on group decisions. The success of his team had always rested on a careful and methodical approach to problem solving, on moving forward one step at a time, and on trying to improve on an existing system rather than striking out into the unknown with a radical innovation. Huntsville was aghast at the risks that the air force and Convair, its prime contractor, had taken in the design of the Atlas. One of the army's rocket engi-

German rocket pioneer Hermann Oberth *(foreground)* with officials of the Army Ballistic Missile Agency in Huntsville, Alabama, 1956. *Clockwise from left:* Ernst Stublinger, Maj. Gen. Holger Toftoy, Wernher von Braun, and Eberhard Rees. (Courtesy of NASA)

neers characterized the new missile as an example of "blimp" construction. Krafft Ehricke, a former member of the von Braun group who had taken a job with Convair, countered that the robust—and heavy—Huntsville rockets represented Brooklyn Bridge construction.

In November 1956, seeking to rationalize the rocket programs operated by the various services, Secretary of Defense Charles E. Wilson announced a new division of responsibility for missile development. The U.S. Air Force would develop the Atlas, Thor, and Titan missiles; the U.S. Navy would proceed with the Polaris project; and the U.S. Army, the only service with extensive missile experience, would be restricted to the operation of rocket weapons with a range of less than 200 miles (320 kilometers).

Secretary Wilson's directive did not mention the Jupiter program. Clearly, however, an army authorized to deploy only short-range rockets had no use for such a weapon—or for the services of the veteran long-range rocket builders at the ABMA. Rumors circulated in Washington and Huntsville that the von Braun team would be disbanded, its members transferred to the other military rocket programs or encouraged to find work in industry.

Two courses of action seemed open to the army. Maj. Gen. John Medaris, the new commander of Redstone Arsenal, which included von Braun's Guided Missile Development Division, argued that on the basis of its experience, personnel, and facilities, the ABMA should be allowed to conduct research and development on long-range missiles for the other services. Several of his subordinates, notably Col. John C. Nickerson Jr., disagreed, suggesting instead that the army launch a full-scale campaign to reverse Wilson's decision.

When Medaris rejected Nickerson's approach, the colonel leaked copies of secret reports on the extraordinary success of the Jupiter program to members of the press, along with his own memorandum charging that Adm. Arthur W. Radford, chairman of the Joint Chiefs of Staff, had plotted the destruction of the army's unique rocket program with air force leaders and representatives of the aerospace industry. When the source of the leak came to light, the full weight of army bureaucracy, from the chief of staff to General Medaris, came crashing down on Nickerson. Court-martialed, fined, and suspended from duty for one year, he resigned his commission. There were those, particularly in Huntsville, who regarded Nickerson as a somewhat flawed martyr. Whatever they might have said to one another in confidence, however, in public the members of the von Braun team stood four-square behind their military superiors, one of the wisest decisions they would ever make.

From the ABMA's point of view this public washing of the Pentagon's dirty linen was a story with a happy ending. "Engine" Charlie Wilson, so named for his long association with General Motors, backed away from his original decision to strip the army of its involvement with long-range rockets, then resigned from office. Neil McElroy, the new secretary of defense, reversed the 200-mile (320-kilometer) limitation that had been placed on army missiles and announced that both the Thor and Jupiter programs would proceed toward production and operational deployment. Once again von Braun and his team had survived.

If Redstone had seemed a natural extension of the V-2, Jupiter represented a relatively small step beyond Redstone. Still, if this was a small step, it was also a critical one. For the first time an American payload, in this case a nuclear warhead, would have to be returned from space. Working with specialists at the Jet Propulsion Laboratory, ABMA engineers developed an ablative heat shield for the warhead that would burn off slowly during reentry, carrying away with it much of the heat generated by atmospheric friction.

In order to test their new recovery system, von Braun and his team needed an existing rocket that was powerful enough to boost an experimental nose cone into space. Engineers at the Rocketdyne Division of North American Aircraft suggested a few simple modifications that would permit Redstone engines to burn a new fuel called Hydyne, which was much more potent than alcohol or kerosene. Moreover, because the test nose cone would weigh considerably less than the nuclear warhead that the Redstone had been designed to lift, the Redstone could be modified with larger tanks, enabling the engine to run for an additional thirty-four seconds.

Finally, the Huntsville team added two upper stages. A small, cylindrical structure, or "tub," mounted on the vehicle's nose would house clusters of eleven second-stage and three third-stage rockets, smaller versions of the JPL's solid-propellant Sergeant. Before launch, the entire tub assembly would be spun by electric motors to provide stability in flight. The tub's second-stage cluster of eleven rockets would fire after the first-stage Redstone booster exhausted its propellants. The ignition of the three rockets in the third stage would boost the payload even higher. In recognition of its contribution to the Jupiter program, the modified Redstone would be known as the Jupiter-C.

Launched from Cape Canaveral, Florida, on 20 September 1956, the pioneer Jupiter-C reached a peak altitude of almost 700 miles (1,120 kilometers) and traveled to an impact point more than 3,000 miles (4,800 kilometers) down range. The second rocket carried a test model of the Jupiter nose cone through a safe reentry, but a guidance malfunction prevented its recovery. The next test flight led to the first successful recovery of an object that had flown into space. The Jupiter-C brought the von Braun team within striking distance of orbital space flight. A quick calculation indicated that the addition of a single fourth-stage Sergeant motor would be sufficient to boost a small nose cone into orbit. Thus modified, the four-stage Jupiter-C would be named Juno 1.

As one successful Jupiter-C test followed another, it occurred to Pentagon officials that the Huntsville team might be sorely tempted to upstage the navy's Vanguard, the official U.S. satellite program, with an "accidental" satellite launch. The Department of Defense (DoD) informed Medaris that he would be held responsible for any such "accident," and ordered him to inspect each missile being prepared for reentry testing to make sure no fourth stage—or possible satellite—was attached.

Had Sergei Korolev been aware of the situation in Huntsville, he might have breathed a bit easier. The first launch of an R-7, scheduled for March 1957, had been canceled because of technical problems. A second attempt in April also was scrubbed before launch. The missile left the pad for the first time on 15 May, only to explode fifty seconds after liftoff. "When things are going badly, I have fewer friends," Korolev told his wife. "My frame of mind is bad, I will not hide it. It is very difficult to get through our failures."[1] Yaraslave Golovanov, a laudatory biographer, admitted that Korolev often had a difficult time accepting responsibility for failures. Some months before the success of Sputnik 1 the chief designer suggested that a launch failure might have been the result of a shortcoming in the propulsion system. "What about you and your rocket," his old rival Valentin Glushko fumed, "are you a saint?"[2]

It was a difficult time. All told, there were eight unsuccessful R-7 launches between May and August 1957, although some of the rockets traveled a considerable distance before exploding or falling out of control. Finally, on 20 August, an R-7 traveled the full length of its range, splashing into the Pacific some four thousand miles (sixty-four hundred kilometers) from the pad. Less than three weeks later, another rocket repeated this performance. All was forgiven. Soviet premier Nikita Khrushchev, who witnessed the September launch, gave his approval for a satellite mission.

Korolev and his allies, including Mikhail Tikhonravov and Mstislav V. Keldysh, director of space sciences for the Soviet Academy of Sciences, decided not to take chances on a large, complex satellite. Instead, despite the fact that the R-7 was capable of lifting more than a ton, they designed a 184-pound (84-kilogram) satellite called Sputnik, or "fellow traveler." Outfitted with only a radio transmitter, this silver ball, twenty-three inches (fifty-eight centimeters) in diameter, was intended to do nothing more than announce its presence in space—loud and clear. It did its job very well.

On the evening of 4 October 1957 news of Sputnik burst like a thunderclap. It was just after six o'clock on the East Coast of the United States when the Soviets announced the successful launch of the world's first artificial Earth satellite. Sputnik 1 had arrived.

In Cambridge, Massachusetts, a small symphony orchestra, its members drawn from the staffs of the Harvard and Smithsonian Astrophysical Observatory, had just begun a rehearsal of Prokofiev's *Peter and the Wolf.* One by one, key officials were called away from their music stands, briefed, and sent off to begin calculating the spacecraft's orbit, setting up a satellite-tracking program, and responding to the calls flooding the switchboard.

In Texas, Senator Lyndon Baines Johnson was throwing one of his legendary barbecues at the LBJ ranch. After dinner, having heard the news, he strolled down toward the Pedernales River with a few of his guests and peered up at the sky. "In the Open West," he later wrote, "you learn to live closely with the sky. It is a part of your life. But now, somehow, in some new way, the sky seemed almost alien. I also remember the profound shock of realizing that it might be possible for another nation to achieve technological superiority over this great country of ours."[3]

In Huntsville, Alabama, the scientists and engineers of the ABMA were at a reception honoring Secretary of Defense Neil McElroy when news of Sputnik arrived. At first Wernher von Braun was stunned, and then almost uncontrollably angry. "We knew they were going to do it," he told the secretary. "For God's sake, turn us loose and let us do something. We can put up a satellite in sixty days, Mr. McElroy! Just give us the green light and sixty days!" Von Braun's superior, Gen. John Medaris, stepped forward. "No, Wernher," he remarked, "ninety days."[4]

The world was never quite the same after 4 October 1957. Human beings had set off on their ultimate voyage of discovery, with the red banner of the Soviet Union leading the way. It was no accident that Sputnik 1 went into orbit just four weeks before the fortieth anniversary of the October Revolution: Sergei Korolev and his comrades had provided the perfect birthday present for a people who had waited four decades for a golden age of socialism that had seemed always to be one five-year plan away.

Euphoria swept the Soviet Union. Sputnik had not shortened the lines for bread and soap, nor otherwise eased the burdens of daily life, but it was something in which Soviet citizens could take genuine pride. They had beaten the West into space. For once it was the Americans who would have to do the catching up. And whatever the future might hold, no one could ever steal this honor from the people of the USSR.

The Soviet achievement hit the Americans like a blow, leaving them stunned and bewildered. Most of them regarded the Cold War as a titanic struggle against the forces of darkness. The task of defeating global communism, they knew, would be enormously difficult, but most were confident that their nation's basic moral strength and the superiority of its technology ultimately would enable them to prevail. Then came Sputnik—and worse. On 3 November 1957 the 1,118-pound (508-kilogram) Sputnik 2 carried the world's first space traveler, a dog named Laika, into orbit. American space planners, who up to a few weeks earlier would have been happy just to get their 3.25-pound (1.47-kilogram) Vanguard 1 into orbit, reeled. Clearly, the rocket that had launched Sputnik 2 also was capable of delivering nuclear weapons on U.S. targets.

The Soviet successes sparked a crisis of confidence in America, something rare in this country's history. Elder statesman and business leader Bernard Baruch characterized the doubts and fears of his fellow citizens with a touch of dark humor. "If America ever crashes," he remarked, "it will be in a two-tone convertible." Senator Styles Bridges of New Hampshire spoke more bluntly: Americans, he said, would have to be less concerned with the "height of the tail fin in the new car and be more prepared to shed blood, sweat, and tears if this country and the free world are to survive."[5] President Dwight D. Eisenhower, himself only mildly surprised by Sputnik, was shocked to discover that it seemed to matter so much to so many. Administration officials did their best to put matters into perspective and restore confidence, but Congress and citizens at large continued to demand action. The Soviets had engaged the United States in a "space race." With national power and prestige at stake and the whole world watching, Americans were eager to take up the challenge.

On the day after the launch of Sputnik 1, von Braun gave Secretary of Defense McElroy a tour of the ABMA's facilities in Huntsville, impressing upon him the fact that the agency had in storage two Jupiter-C launchers, which, within ninety days of an order to proceed, could be modified into Juno 1 vehicles and used to launch a satellite. McElroy remained silent. "When you get back to Washington and all hell breaks loose," von Braun advised him, "tell them we've got the hardware down here to put up a satellite any time."[6]

On 8 November 1957, five days after the launch of Sputnik 2, Gen. John Medaris was ordered to "prepare" to orbit a satellite. Convinced that the powers that be in Washington still intended to hold their team in reserve until after the first Vanguard attempt, both Medaris and von Braun threatened to resign unless they received permission to launch as soon as their preparations were complete. Their orders were modified to that effect; Wernher von Braun was back in the space business.

While the 4,735 employees of the ABMA set to work on the launch vehicle, William Pickering and the staff of the JPL assembled the satellite. Explorer 1, as the little craft was called, would measure 80 inches (200 centimeters) long and 6.5 inches (16.5 centimeters) in diameter and weigh only 30.7 pounds (13.9 kilograms). Unlike Sputnik 1, the U.S. satellite would carry four sets of instruments, as well as a telemetry system that would broadcast important scientific information back to Earth. Two of the instrument packages would detect the impact of micrometeoroids, and four gauges would record both interior and exterior temperatures.

The most important experiment, designed to measure the level of cosmic radiation, was the work of physics professor and veteran space scientist James Van Allen and Wei Ching Lin, one of his graduate students at the University of Iowa. Questions relating to cosmic radiation had intrigued U.S. scientists since the late nineteenth century. How much energy did the Sun shower upon the earth? How did this radiation interact with the earth's magnetic field and atmosphere? Was there a connection between cosmic-radiation levels and changes in the earth's climate?

Long before the precise nature of cosmic radiation was understood, physicists and astronomers had been climbing mountains and sending aloft balloon-born instruments in an attempt to measure solar energy free of the atmosphere's filtering effects. Continuing this line of research, the Smithsonian's Charles G. Abbot had sponsored the work of rocketeer Robert Goddard in hopes of obtaining a device that could lob his instruments beyond the atmosphere.

James Van Allen was now about to realize Abbot's dream. Having spent years piecing together information on cosmic radiation in both the upper atmosphere and near-Earth space that had been obtained from dozens of sounding-rocket and balloon flights, he was more than ready for the overview that a satellite could provide. He had been delighted to build a small radiation sensor and transmitter for the Vanguard program, and he was happy to prepare a second for Explorer 1.

By the late fall of 1957 excitement ran high at Cape Canaveral, where two rocket teams raced to join the Soviets in space. The U.S. Navy/Martin engineers were first off the pad—barely. At 11:45 A.M. on 6 December 1957 the Vanguard TV-3 rocket rose less than five feet into the air, sank back down, and toppled over in a cloud of smoke and flame. An enormous explosion rocked the complex as the propellant tanks burst. Only one thing seems to have worked as planned that day: when the firefighters finally were able to move into the area, they found the little silver ball of a satellite, charred and dented, faithfully beep-beep-beeping away on the ground.

America's spirits hit rock bottom with the loss of the Vanguard TV-3, and the pressures on the Huntsville team grew accordingly. Their hopes rested on a Juno 1 that was already in the hands of Kurt Debus, head of the ABMA Missile Firing Laboratory at Cape Canaveral. No one knew more about launching rockets than Debus and his crew. Twenty-year veterans of the rocket-building trade, they had conducted hundreds of launches at Peenemünde, White Sands, and, most recently, Cape Canaveral.

The Juno 1's launch date, originally set for 29 January 1958, was scrubbed two days in a row because of high winds aloft. Finally, at 10:55 P.M. on the thirty-first, the moment of truth arrived, and the rocket roared skyward from its metal launch stand.

Doppler radar tracked the vehicle's ascent. If all went well, the Juno 1, with its Explorer 1 payload, would be coasting to its peak altitude four hundred seconds into the flight. The tub carrying the three top stages would separate from the rocket's main body. Based on the tracking data, Ernst Stuhlinger, von Braun's chief theoretician, would calculate the apex of the tub's trajectory and push a button igniting the eleven second-stage boosters at precisely that moment. Guidance equipment attached to the tub would set the correct

The Vanguard TV-3 rocket, carrying a satellite intended for Earth orbit, explodes on the pad. The missile was destroyed, but the satellite, dented and burned, was found in working condition amid the wreckage. (Courtesy of NASA)

angle for the second stage, but there was no such equipment to guide the third and fourth stages, the last actually part of the satellite, when they exited the spinning tub.

Von Braun was at the Pentagon, preparing to join Secretary of the Army Wilbur Brucker, James Van Allen, and JPL's William Pickering at a press conference if all went well. Just after 11:00 P.M. von Braun received a call from the cape and was informed of the exact launch time. Pulling out a pad and pencil, he made a few quick calculations and announced to Brucker and the others that they should receive confirmation of success when the satellite passed over a San Diego tracking station at precisely 12:41 A.M. Pickering established an open line to the tracking center.

But 12:41 came and went with no sign of Explorer 1. The tension built as the seconds ticked by. After five minutes with no word, von Braun began to wonder how he was going to explain what had gone wrong. Finally, at 12:49 A.M., a voice on the other end of the phone line told Pickering that the tracking station had picked up the satellite. The orbit simply had ended up a bit higher than predicted. With the crisis passed, von Braun, Van Allen, Pickering, and their colleagues drove across the Potomac for an early morning press briefing at the National Academy of Sciences. That afternoon pictures of the three smiling men holding a full-scale model of Explorer 1 over their heads in triumph appeared in papers from coast to coast.

The Soviet Union had led the way into space, but the United States scored the first great scientific breakthrough of the era. Explorer 1's radiation counters discovered rings of charged particles trapped in the earth's magnetic field. Vanguard 1, which was successfully launched on 17 March 1958, confirmed the existence of what would be called, fittingly, the Van Allen radiation belts.

On 26 March the ABMA team sent up Explorer 3, the world's fifth satellite. The Soviets countered on 15 May by orbiting the 2,925-pound (1,316-kilogram) Sputnik 3 geophysical satellite. The United States would have to "launch very many satellites the size of oranges," Nikita Khrushchev commented, "in order to catch up with the Soviet Union."[7] Undaunted by the small size of their payloads, or by the words of the Soviet premier, the ABMA rocketeers forged ahead with the launch of another scientific satellite, Explorer 4, on 26 July 1958. Medaris and von Braun had achieved Holger Toftoy's goal of a successful U.S. space program that operated under the aus-

William Pickering, director of the Jet Propulsion Laboratory *(left);* **James Van Allen of the University of Iowa** *(center);* **and Wernher von Braun** *(right)* **proudly display Explorer 1, the first successful U.S. satellite. Launched on 31 January 1958, Explorer 1 made the first great discovery of the space age: the rings of charged particles trapped in Earth's magnetic field that would be named for Van Allen. (Courtesy of NASA)**

pices of the U.S. Army, and they were more than prepared to lead on into the future.

A month before the launch of Explorer 1, Medaris and von Braun had presented their Pentagon supervisors with an ambitious fourteen-year plan for the conquest of space. Given a free rein and sufficient funding, they proposed to develop a family of large boosters, send human beings around the Moon in 1963, build a space station, and establish a large and permanent lunar colony by 1971. It was the *Collier's* series come to life.

The intraservice rivalries that had marred the U.S. space effort since 1945 remained at work, however. In the fall of 1957, Herblock, the most influential political cartoonist of the era, delighted readers of the *Washington Post* with a drawing of two beribboned generals mopping their brows in relief as a rocket bearing the hammer and sickle of the Soviet Union roared overhead. "Whew!" one officer gasped, "At first I thought it was sent up by one of the other services."[8] Indeed, the air force and navy were quick to contest the idea of a U.S. space program dominated by the army. Both services could point to their long involvement with research-airplane and manned-balloon programs that had extended the limits of speed and altitude and helped blaze a path toward space flight. The navy had sponsored the Viking and Vanguard projects and had pioneered innovative management techniques to coordinate the development of a fleet of nuclear-powered submarines armed with Polaris missiles. Air force publicists, meanwhile, began to focus on USAF operations in the "aerospace" environment, emphasizing in particular the role that their new research craft, the North American X-15, would play in the transition to flight operations outside the effective limits of the atmosphere. They also promoted the air force's success in managing the complex Atlas, Titan, and Minuteman rocket programs, each of which would result in a rocket far more powerful than anything Huntsville had produced to date.

In the wake of the Sputnik crisis Senate majority leader Lyndon B. Johnson held hearings before the Special Committee on Science and Astronautics to identify problems in and solicit recommendations for the U.S. space effort. Johnson called scientists, engineers, defense analysts, and military officers before the panel, and virtually all of them suggested improved planning and coordination. How this might be achieved, though, was another question.

Naturally, each of the three military services volunteered to take command of a unified space program. Other authorities suggested that the space effort might be organized by an existing organization, such as the Atomic Energy Commission, the National Science Foundation, the National Academy of Sciences, the National Advisory Committee for Aeronautics (NACA), or by an entirely new agency within the DoD. Senator Mike Mansfield of Montana proposed the creation of a cabinet-level Department of Science.

President Dwight Eisenhower faced a dilemma. The need for federal sponsorship of research-and-development programs was clear. Science and technology were the basis for military security, the very bedrock of national power and prestige in the twentieth century. At the same time, the economic well-being of the nation—and the direction of its public policy—rested in the hands of those scientists, engineers, government officials, and industrial leaders who commanded the forces of science and technology. "We must, be alert," he warned his fellow citizens, "to the . . . danger that public policy could itself become the captive of a scientific-technological elite."[9]

Eisenhower and his advisers realized there were no easy choices. Nor, following the launch of Sputnik 2, could they avoid dealing with the problem of organizing a coherent U.S. space program. They insisted on one important guideline: although such key aspects of the national space effort as the development of reconnaissance satellites would remain in the hands of the military, the bulk of the U.S. space program would remain under civilian control.

Presidential science adviser James Killian and the members of the President's Science Advisory Committee favored the transformation of the venerable NACA into a new space agency. NACA had been established by President Woodrow Wilson on 3 March 1915 "to supervise and direct the scientific study of the problems of flight, with a view to their practical solutions." It represented a new kind of federal agency, one designed to conduct basic research that would advance a technology regarded as essential to the nation's defense.

The men and women of NACA had spent the subsequent forty years working quietly and with little fanfare. The agency's Langley, Ames, and Lewis laboratories in Virginia, California, and Ohio, respectively, contributed to a revolution in flight technology that transformed the airplane from a thing of wood, wire, and fabric into a sleek, streamlined, all-metal craft capable of carrying passengers and freight across continents and oceans or delivering death and destruction from the sky. As one British engineering journal stated, "The present-day American position [of preeminence] in all branches of aeronautical knowledge can, without doubt, be attributed to this far-seeing [NACA] policy."

During the years after World War II, NACA's leaders had focused their attention on the problems encountered by aircraft flying at extreme speeds and altitudes. The sky itself became a laboratory. Members of the Pilotless Aircraft Research Division pursued their studies at an abandoned navy test facility at Wallops Island on the Eastern Shore of Virginia, about one hundred miles north of the Langley Research Center in Hampton. Here they dropped small test shapes from high-flying aircraft and fired them into the atmosphere aboard rockets, recovering vital information about flight characteristics via telemetry broadcast from the models' sensors.

Out at Muroc Dry Lake in the high California desert, NACA and the military services established a vast area for testing the world's most advanced aircraft. It was here, on 14 October 1947, that Capt. Charles "Chuck" Yeager, piloting the Bell X-1, became the first man to fly faster than sound. Over the next decade, navy, air force, and NACA researchers pioneered new technologies and probed the limits of flight with other experimental aircraft in this series.

Space flight did not come as a surprise to NACA, but most staff engineers and scientists thought that flight beyond the atmosphere would be achieved gradually. By the spring of 1958, however, NACA was expending roughly one-fifth of its time and energy on problems directly related to space flight. The decision to transform NACA into a space agency made much sense. The organization's experience in finding solutions to the technical problems of flight would be of obvious value. Moreover, NACA had a history of successful cooperation with both the military and industry.

On 29 July 1958 President Eisenhower signed into law the National Aeronautics and Space Act, thus creating, on NACA's foundation, the National Aeronautics and Space Administration—NASA. In August Eisenhower appointed T. Keith Glennan, president of the Case Institute of Technology and a former commissioner of the Atomic Energy Commission, as NASA's first administrator. Hugh Dryden, NACA's last director, would remain as the new agency's deputy administrator.

NASA began its work with enthusiastic support from administration officials, congressional leaders, and the Pentagon. Still, it faced a host of problems. Unlike NACA, which basically had been a research and problem-solving organization, NASA would be responsible for every aspect of the nation's civilian space program, including contracting with the aerospace industry for launch vehicles, satellites, and manned spacecraft; constructing launch and tracking facilities; conducting the research needed to support those programs; and hiring personnel to launch, monitor, and fly space missions. No longer in the business of supporting military research-and-development programs as NACA had been, NASA quickly would become something of a competitor.

On 1 October 1958 Glennan and Dryden gathered NASA's headquarters staff together in the courtyard of their building, the Dolly Madison House, on Lafayette Square near the White House, to celebrate the end of one era and the beginning of another. NACA, which had served as the world's most successful aeronautical research-and-development agency, was no more. In its place stood a new and untried organization. Ready or not, the National Aeronautics and Space Administration was in business.

9 • This New Ocean

The National Aeronautics and Space Administration was born of the frustration and uncertainty of the Sputnik era. From the outset great things were expected of the men and women who worked for NASA, including what most people a few years earlier had regarded as impossible: to fly into space. And they would be expected to do it with the whole world looking over their shoulders.

Hugh Latimer Dryden, a brilliant engineer who had earned a reputation as one of the most competent administrators in Washington, spearheaded the conversion of NACA to NASA between March and August 1958. An eleven-year NACA veteran, Dryden had assumed that the vigorous pursuit of space science would be NASA's major goal. He prepared a series of preliminary organization charts, each of which included an office of space science headed by a deputy administrator reporting directly to the administrator. He also cooperated with Detlev Bronk of the National Academy of Sciences, Alan T. Waterman of the National Science Foundation, Herbert York of the DoD's Advanced Research Projects Agency (ARPA), and Lloyd Berkner, president of the International Council of Scientific Unions to develop a replacement for the old Upper Atmosphere Research Panel. Together they created the Space Science Board, which would coordinate efforts in this field and advise the new agency on matters relating to research.

T. Keith Glennan, who was sworn in as NASA's administrator on 19 August 1958, took a different view. A former president of the Case Institute of

Technology and a member of the Atomic Energy Commission, he recognized that scientific research and the pursuit of practical benefits from space flight would always be a part of NASA's program. But he had no doubt as to the agency's primary goal: NASA was in business to overtake the Soviet space effort and restore confidence in the superiority of U.S. technology. For a time at least, the performance of NASA's machines would be more important than the information they returned.

In Glennan's organizational chart, an assistant director for space science was subordinate to a director of space flight programs, whose primary concern would be the manned space-flight effort. The first generation of NASA space scientists would work under the direction of aeronautical engineers who were also experienced project managers, having cut their professional teeth with the old NACA. Some scientists feared that their goals would become secondary to the engineers' desire for a smooth mission.

Abe Silverstein, deputy director of the Lewis Flight Propulsion Laboratory, in Cleveland, Ohio, and an old NACA hand, was named first director of the Space Flight Development Office, with Homer Newell serving as his assistant director in the Office of Space Science and Applications. Newell immediately began to forge what he would characterize as a love-hate relationship with the Space Science Board. The board's advice would be welcome, but NASA would have the final word in the creation of an integrated national program for the scientific exploration of space. The Upper Atmosphere Research Panel's absolute control over space research was at an end. Newell's relationship with the board paralleled NASA's relationship with the space-science community in general.

NASA inherited all of NACA's facilities, including the Langley laboratory and Wallops Island test station in Virginia, the Lewis laboratory in Ohio, and the Ames laboratory and Muroc flight station in California. Congress also authorized the construction of a third major laboratory, the Goddard Space Flight Center, in Greenbelt, Maryland, which opened in 1961. In addition, President Eisenhower transferred to NASA virtually all of the existing non-military space projects, including the navy's Project Vanguard satellite program, all of the satellite efforts underway at Huntsville, and plans for a lunar probe and a single-chamber rocket engine capable of producing 1 million pounds of thrust (4,448,399 newtons), which was being promoted by the newly formed ARPA.

To avoid any delay in critical programs Glennan asked the DoD to retain temporary control of most of the projects in question while he organized his own agency, a process that would include drawing even more existing organizations into the NASA circle. In October 1958 the new administrator asked for and received permission to take control of operations at the Jet Propulsion Lab in California.

Glennan also visited Huntsville and decided that although the von Braun team was fully occupied with missile projects, it was really interested in the Moon. Glennan offered the scientists an opportunity to go to work for him on the one-million-pound-thrust engine program that he had inherited from von Braun via ARPA. Von Braun's loyalty to Medaris and the ABMA overcame his desire to dedicate the energies of his team to a pure space program, but the episode enhanced Glennan's reputation as something of a pirate.

By September 1959, however, the situation had changed. Secretary of Defense Neil McElroy had assigned all military space projects to the air force and suggested to Glennan that the time was now ripe to fold von Braun's operation into NASA. Beaten at last, General Medaris lodged a strong but ultimately fruitless protest and retired soon thereafter. On 1 July 1960 a significant portion of the ABMA operation in Huntsville became NASA's George C. Marshall Space Flight Center. For the first time in his long career, Wernher von Braun had a civilian boss and a clear directive to forge ahead with the development of a powerful booster designed solely to carry a payload into space.

Von Braun's superbooster would be the capstone of a new generation of NASA space-launch vehicles, including Scout, a solid-propellant booster designed to loft 130-pound (59-kilogram) payloads into low-Earth orbit, and Centaur, a high-energy upper stage that was fueled by liquid hydrogen and capable of boosting robot probes into deep space. Until those rockets were ready for flight, NASA had to make do with its stable of Vanguard, Juno, Redstone, Thor, and Atlas missile/launch vehicles.

Structuring NASA's overall space-flight program and selecting experiments to be flown aboard prospective satellites presented the new agency with some interesting challenges. Its first operational missions involved attempted launches of the Explorer, Vanguard, Beacon, and Pioneer spacecraft inherited from the ABMA, navy, and ARPA—relatively simple vehicles that were intended to study cosmic radiation and extend the exploration of the magnetosphere begun by the United States' first two satellites.

The Soviets, meanwhile, had moved on to bigger and better things. Having written themselves into the history books with the early Sputniks, chief designer Sergei Korolev and his team targeted the Moon as their next goal. They proceeded to launch the first payload to escape Earth's gravity (Luna 1), the first spacecraft to collide with another world (Luna 2), and the first spacecraft to obtain photographs of the far side of the Moon (Luna 3). Moreover, the Soviets accomplished their goals without developing an entire stable of launch vehicles: the SL-3 (the Western designation for the 1959 variant of the R-7) had proved adaptable to every need. The Soviet Union was first in space, and it clearly intended to retain that position.

NASA got off to a slow and disappointing start. All four of its attempts to orbit payloads in 1958 ended in failure. The record for 1959 stood at nine successes for fourteen tries; in 1960 seventeen launch attempts yielded twelve successes. The first three Pioneer spacecraft, with which NASA hoped to race the Soviets to the Moon, all failed. But NASA took an unequivocal lead in the application of space technology to the solution of problems on Earth. Its eighteen successful launches of 1959–60 included such impressive achievements as the first photo of the Earth from space (Vanguard 2, 7 February 1959), the first successful weather satellite (TIROS 1, 1 April 1960), the first navigation satellite (Transit 1B, 13 April 1960), and the first passive communications satellite (Echo 1, 19 August 1960).

The revolution in space science and applications-satellite technology occurred with little or no public fanfare. By the early 1960s the curtain was about to go up on the next act in the cosmic drama of the space age. Most citizens of Planet Earth were reserving their attention, enthusiasm, and applause for the first human beings to venture beyond the atmosphere. Korolev and his colleagues had every intention of sending the first person into space. U.S. space planners were just as determined to overtake their rivals and restore world confidence in American know-how. Von Braun's carefully prepared blueprint for space flight went straight out the window: there would be no time for a systematic, step-by-step approach to a permanent human presence in space. The operative phrase was "man in space soonest," or MISS.

Immediately after the 4 October 1957 launch of Sputnik 1, the military services had rushed forward with competing proposals for a manned space program. The army called for a minimum suborbital effort known as Project Adam, which entailed a quick trip up into space and back that NACA's Hugh Dryden compared to " shooting a young lady from a cannon." The navy proposed a much more complex scheme that included returning an astronaut to

Earth aboard an inflatable glider. Air force officials promised to launch a single-seat spacecraft aboard a modified Atlas missile.[1]

The best approach to manned space flight was also a matter of heated debate within NACA. Traditionalists argued for a new kind of research aircraft that could take off from a runway, fly into space, and glide back to a landing on Earth. By 1957 such a scheme was no longer a pipedream. The XLR-99 rocket engine designed to power the North American/NACA X-15 could develop 57,000 pounds of thrust (253,558 newtons), just short of the 76,000 pounds (338,078 newtons) generated by a Redstone booster. The X-15 could reach an altitude of 67 miles (108 kilometers) and achieve a maximum speed of 4,532 miles (7,297 kilometers) per hour. By contrast, the first suborbital Mercury flights of 1961 would reach a speed of only 5,180 mph (8,340 kph).

The problem was that the X-15 could not achieve orbital velocity of some 18,000 mph (28,980 kph), nor could it safely reenter the atmosphere from orbit. An aircraft capable of that level of performance remained a generation or two away. H. Julian Allen of the Ames Research Center explained the problem to his NACA colleagues in the clearest possible terms. Although such a plane might be the safest and surest way to send human beings into orbit, it was by no means the fastest. If the goal was to put a man in space soonest, engineers would have to abandon conventional aerodynamic thinking in favor of something like the army–air force "cannonball" schemes. Allen suggested using an existing missile to launch into orbit a blunt, possibly saucer-shaped spacecraft. The less streamlined the better, Allen explained. The idea was to rely on the capsule's blunt design to create enough air resistance to slow its speed during reentry.

Inspired by Allen's call for new ideas, NACA engineer Maxime Faget proposed a small, cone-shaped spacecraft that was light enough to be launched on suborbital flights by a Redstone or to be sent into orbit by the more powerful Atlas missile. When the time came to return the craft from space, a small cluster of rockets on the vehicle's blunt end would be fired in the direction of its motion to slow its speed and cause it to fall out of orbit. The craft then would reenter the atmosphere blunt end first, and parachutes would lower it down to a landing in the ocean.

Problems had to be resolved. The atmospheric friction that slowed the craft also would generate more than enough heat to vaporize it. But Faget was confident that such difficulties could be overcome, and, by 29 July 1958, when President Eisenhower signed the National Aeronautics and Space Act, a plan to send human beings into space in a relatively short period of time was in place. Charged with the development and operation of vehicles capable of carrying instruments, equipment, supplies, and living organisms through space, NASA, not the army, navy, or air force, would be sending the first Americans into orbit.

Several months before NASA opened its doors for business Faget and a group of his colleagues set to work transforming the Allen-Faget concept into the detailed design for a spacecraft. A titanium pressure vessel that was just large enough to house a single occupant would be surrounded by a conical outer shell of protective nickel-steel-alloy shingles designed to radiate heat away from the spacecraft. A heat-resistant beryllium canister on the nose held the parachutes that would slow the craft's fall into the ocean at the conclusion of the mission.

From the point of view of a prospective astronaut, the convex heat shield fitted to the blunt base of the craft was the single most important feature of the design. When subjected to extreme temperatures, the resins in the plastic and fiberglass shield would give off gases, dissipating heat and protecting the vehicle during its long plunge back into the atmosphere. A cluster of three solid-propellant rockets strapped to the shield would be fired to initiate reentry and then jettisoned.

NASA was less than a week old when administrator Glennan approved the Faget plan and created a Space Task Group (STG) of thirty-five Langley engineers, headed by Robert Gilruth, to push the work forward. On 7 November 1958 members of the STG convened to discuss the specifications for their spacecraft and invited forty aerospace firms to bid on its construction. On 17 December, fifty-five years to the day after the Wright brothers had flown at Kitty Hawk, NASA officials publicly announced the details of Project Mercury, the U.S. manned space program. Less than a month later, on 12 January 1959, agency officials revealed that McDonnell Aircraft of St. Louis, Missouri, had won the competition and would serve as the prime contractor for the Mercury spacecraft.

The selection of the human beings who would ride the Mercury spacecraft into orbit was complete as well. The astronauts had passed through an extraordinary screening process. "What we're looking for," one air force general remarked, "is a group of ordinary Supermen."[2] At the outset, some federal officials had argued that the first space travelers should be chosen from the ranks of daredevils and athletes: racecar drivers, high divers, circus performers, and the like—men (no women were considered) accustomed to taking great risks. That proposal did not seem appropriate to President Eisenhower, who felt that applicants should be selected from the existing pool of trained military test pilots.

An initial search turned up 508 possible candidates. A winnowing process that included the study of personnel records, comments of commanding officers, and personal interviews finally reduced that number to 32 test pilots. Each of these men was younger than forty, weighed less than 180 pounds (81 kilos), stood less than 5 feet 11 inches (180 centimeters) tall, and held an engineering degree. The finalists were probed, prodded, and subjected to batteries of medical and psychological tests. In the end, seven were chosen.

On 9 April 1959 NASA introduced the seven Mercury astronauts to the world: air force officers (all captains) Donald K. "Deke" Slayton, Virgil I. "Gus" Grissom, and L. Gordon Cooper; naval aviators (all lieutenant commanders) Alan B. Shepard Jr., Walter M. Schirra Jr., and Malcolm Scott Carpenter; and a lone marine, Lt. Col. John H. Glenn Jr. Most of them were small-town boys. All were married with children. Only four inches separated the tallest (Alan Shepard at 5 feet 11 inches, or 180 centimeters) from the shortest (Gus Grissom at 5 feet 7 inches, or 170 centimeters). Wally Schirra, who had to drop five pounds from his normal 185 (83 kilos), was the heaviest. Gordon Cooper, who weighed in at 150 pounds (68 kilos), was the lightest— and, at age thirty-two, the youngest. Thirty-seven-year-old John Glenn was the group's oldest member. The ebullient Schirra was the only one who did not wear his hair in a crew cut.

If the astronauts had much in common, they also were individuals with varied levels of experience. Glenn had the most flying hours and was the most decorated. A veteran of World War II and Korea, he had won the Distinguished Flying Cross five times and wore eighteen clusters on his Air Medal. At the time of his selection he held the coast-to-coast speed record. Both Grissom and Schirra also wore the Distinguished Flying Cross and the Air Medal with clusters. Scott Carpenter, although a graduate of the Navy Test Pilot School at Patuxent, Maryland, had logged only three hundred hours in jets at the time of his selection.

Deke Slayton, a pilot's pilot who was once described as looking like a character out of Steve Canyon, was a shy man who hated public appearances. The first Project Mercury press conference, he remarked, was "the worst stress test I've ever been through. If I hadn't already passed my physicals,"

The Mercury Seven dressed in business attire for a formal portrait. *Front row, left to right:* Walter M. Schirra Jr., Donald K. Slayton, John H. Glenn Jr., and Scott Carpenter. *Back row, left to right:* Alan B. Shepard Jr., Virgil I. "Gus" Grissom, and L. Gordon Cooper. (Courtesy of NASA)

he added, "they would probably have flunked me right then, because my knees were knocking."[3] Schirra and Shepard had well-earned reputations as pranksters. Shepard delighted in mimicking José Jimenez, comic Bill Dana's reluctant-astronaut character. Years later, as commander of the Apollo 14 lunar mission, he surprised ground controllers by attaching the head of a golf club to a lunar tool handle and teeing off on the Moon. Schirra was a practical joker who could find humor in any situation. Both men loved fast cars: Shepard drove a spotless white Corvette, Schirra a canary-yellow Austin Healy.

Four of the seven astronauts—Carpenter, Schirra, Shepard, and Slayton— were smokers. Each of them quit soon after being selected but eventually took up the habit again. Cooper and Grissom both loved to hunt and fish. Scott Carpenter was a platform diver, an acrobat, and an accomplished ballroom dancer. Gordon Cooper, the son of a World War I aviator, was the only member of the group who owned and flew his own airplane.

The astronauts were instant heroes. Like knights of old, they were seen as champions, men who would risk their lives to carry their nation's banner in a race against the Soviets. Over the next few years their names and faces would become familiar to all Americans. These were the men who would lay to rest any doubts about America's strength and determination and would restore a sense of national pride and confidence.

Ironically, the precise duties of an astronaut were not entirely clear at the time of the selection. "Although the entire satellite operation will be possible, in the early phases, without the presence of man," one NASA source reported, "the astronaut will play an important role during the flight." Small wonder that the astronauts should express some concern about the nature of their responsibilities on a mission that could be performed "without the presence of a man," particularly when it was announced that chimpanzees would take the first rides into space aboard the Mercury capsule. A joke circulating within the test-flying fraternity suggested that the astronauts would have to clean the chimps' droppings off the seat before climbing into the spacecraft; they would be "Spam in the can," passengers who were simply going along for the ride. Determined to counter these jokes and to ensure that as pilots they would exercise full control over their spacecraft, the astronauts used their increasing prestige and public visibility to force essential design changes. At their insistence, tiny portholes gave way to a window; an emergency hatch

The Mercury astronauts take a break from desert survival training. (Courtesy of NASA)

was installed for quick escapes; and the location and function of every control, switch, lever, and instrument was studied and modified as required.

Difficulties with the launch systems underscored the dangers of the enterprise. The Atlas rocket that eventually would boost the astronauts into orbit was plagued by developmental problems. Even the tried and true Redstone was still experiencing difficulties. In November 1960, during the first complete flight test of the Mercury Redstone system at Cape Canaveral, the rocket rose only six inches (fifteen centimeters) off the ground before an automatic engine shutdown settled it back onto the pad with the parachute deployed.

The first flight test of the Mercury Redstone combination with an animal on board did not do much to boost astronaut confidence either. The chimpanzee Ham made a suborbital trip into space on 31 January 1961, blasting off after a series of malfunctions had kept the vehicle and its occupant on the pad three hours longer than planned. Once off the ground, the Redstone subjected Ham to a much faster acceleration and to higher gravitational forces than predicted. A leaky valve led to a severe drop in cabin pressure. The

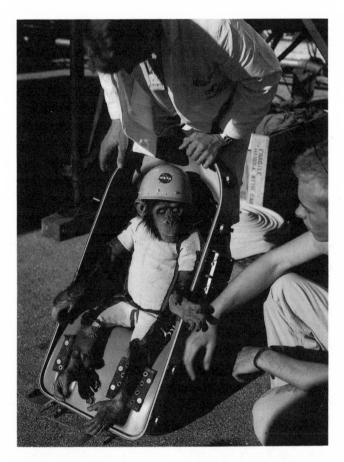

NASA shows off its chimpanzee Ham in training for his suborbital trip into space. Although the 31 January 1961 launch was
three hours late and splashed down 130 miles off target, Ham's
flight was a success—he survived. (Courtesy of NASA)

spacecraft finally splashed down 130 miles (209 kilometers) from the target
area, and a frightened chimp waited for his delayed pickup as water leaked
into the spacecraft through the damaged valve.

With the whole world watching, safety was NASA's first priority. As a result, the first scheduled manned launch was shifted from March to late April
1961 in order to allow for changes and additional testing following the nearly
disastrous Ham flight. Thus, on 12 April 1961 it was Maj. Yuri Alexeyevich
Gagarin, a twenty-seven-year-old Red Air Force pilot, who became the first

person to fly in space. His vehicle, Vostok ("east"), weighed three times as much as a Mercury spacecraft, and he completed a full orbit of the earth, a much more impressive achievement than the suborbital trips planned for the first few manned Mercury flights.

Like Sputnik, the Gagarin flight took the world by surprise. One of the keys to the enormous propaganda value of the Soviet space effort was the fact that it revealed so few details. The whole world witnessed NASA's successes and failures on television, and anyone who picked up a daily newspaper or a trade journal could gain access to detailed information on U.S. rockets and spacecraft. The size and configuration of the Soviet's Sputnik rocket, on the other hand, let alone its technical details, remained a mystery for months after the satellite's launch. Wernher von Braun had become a renowned engineer-scientist, whereas the identity of Soviet "chief designer" Sergei Korolev was completely unknown.

Nikita Khrushchev built his foreign policy on Korolev's successes. "Of course, we tried to derive the maximum political advantage from the fact that we were the first to launch our rockets into space," he wrote in his memoirs. "We wanted to exert pressure on the American militarists—and also influence the minds of more reasonable politicians—so that the United States would start treating us better."[4] The Soviet premier showed his appreciation for Korolev by raising the engineer beyond the reach of his rivals. By 1960 the chief designer and his allies commanded a vast complex of design bureaus, research laboratories, manufacturing plants, and launch complexes, as well as a growing city devoted entirely to the conquest of the cosmos. But this power and prestige came at a considerable price. Khrushchev had developed a ravenous appetite for space triumphs with which to goad the West, triumphs that were increasingly difficult to conjure.

Planning for the Soviet effort to launch human beings into space began at a meeting in the Academy of Sciences building in Moscow early in 1959. Korolev was there, as were Mstislav Keldysh, a leading academician serving as space-programs head; Lt. Gen. Nikolay Kamenin, who would command the cosmonaut corps; Col. Yevgeny Karpov, a flight surgeon who would head the cosmonaut-selection team; and a handful of other space authorities. By July 1959 Karpov and his colleagues had begun visiting Red Air Force bases, seeking pilots who, among other criteria, had suffered in-flight emergencies and had elected to nurse their aircraft home rather than eject.

In time several hundred candidates were ordered to report to the Red Air

Cosmonaut Yuri Gagarin *(left),* the first person to fly in space, with Sergei Korolev, head of the Russian space program. Cosmonauts were chosen from a list of military pilots who had experienced in-flight emergencies and brought their planes down safely rather than ejecting. (Courtesy of the National Air and Space Museum, Smithsonian Institution, Photo No. 73-379)

Force facility at the M. V. Frunze Central Airfield in Moscow for medical and psychological examinations. Their experiences, and their attitude toward flight surgeons, would have sounded familiar to the Mercury Seven. The pool of candidates eventually was winnowed to forty, and then to twenty-nine. Finally, twenty pilots were selected as cosmonauts. Of these, twelve would fly into space, seven would be dropped from the program during training, and one would die in a training accident.

The members of the cosmonaut team were younger than their American counterparts. The eldest, Pavel Belyayev, was thirty-four, whereas Gherman Titov and Georgi Shonin were only twenty-four at the time of their selection. All but one of the cosmonauts was a graduate engineer. Like so many Red Air Force pilots, a number of the cosmonauts initially had been trained as machinists or factory workers. Discipline was apparently something of a problem among the cosmonaut candidates. Three of them were dismissed from the program as a result of a drunken brawl with military police at a railroad station.

On the other hand, the cosmonauts seem to have enjoyed a much keener interest in the arts than did the original astronauts. Both Alexei Leonov and Pavel Belyayev were accomplished amateur painters. Belyayev was also something of a poet, and Titov was fond of quoting long snatches of poetry from memory. Georgi Shonin was a voracious reader who particularly admired the work of the French aviator-author Antoine de Saint-Exupéry. In the end several of the original cosmonauts wrote books on space flight. Yevgeny Khrunov wrote science fiction, and Vladimir Komarov edited books in his spare time. But there were similarities between the two groups as well. Like Wally Schirra, Alexei Leonov struggled with a weight problem. Andrian Nikolayev would have enjoyed nothing more than joining Gus Grissom and Gordo Cooper on a fishing expedition. Valery Bykovsky shared John Glenn's interest in politics, and both Pavel Popovich and Scott Carpenter became interested in UFOs.

Like his American counterparts, Sergei Korolev had to overcome a host of technical difficulties before he could send a human being into orbit. One major problem was the fact that although the trusty Semyorka remained a fine launch vehicle, it left much to be desired as an ICBM. Soviet leaders boasted that ICBMs were rolling off the assembly lines "like sausages." In fact, the R-7, as the Semyorka was designated, had been judged too unreliable for deployment as an operational weapons system. The addition of powerful upper-stage motors to the R-7 had permitted the launch of the three Luna spacecraft, but the need for a new and more powerful booster was apparent. For the time being, however, Korolev had to make do with the materials at hand.

In the case of Vostok, Korolev ran neck-and-neck with the Americans until the last minute. Between 15 May 1960 and 25 March 1961 the Soviets launched a total of five Korabl Sputniks, unmanned versions of the Vostok ship. Weighing five tons (4,500 kilograms) apiece, the spherical capsules were designed to test Vostok recovery techniques and life-support systems. The dogs Belka and Strelka, along with forty mice, two rats, several hundred insects, and numerous plant specimens, survived seventeen orbits aboard Korabl Sputnik 2. Later flights included guinea pigs, frogs, and mannequins on the passenger manifest.

The Korabl Sputnik program was impressive, but it was far from perfect. Launched on 15 May 1960, the original vehicle was incorrectly positioned for retrofire and was lost during reentry. This failure marked the beginning of a year in which the Soviet space program would enjoy great successes—and suffer even greater tragedies.

On 19 September 1960 Khrushchev arrived in New York to attend a session of the United Nations General Assembly. Safely packed away in his baggage was a model of a Mars spacecraft, ready for presentation to Secretary General Dag Hammerskjold once the Soviet premier received word of a successful launch. As usual on the occasion of a politically significant flight, Korolev had prepared a backup launch vehicle.

But there were unexpected problems. In New York, Khrushchev extended his stay, becoming so agitated during one UN session that he pounded his shoe upon the table. Finally, on 10 October, Korolev's team succeeded in launching Mars 1. The Mars version of the R-7 booster functioned perfectly, but the new fourth-stage rocket motor that was to have propelled the spacecraft toward the Red Planet failed to ignite. The next day Khrushchev left New York with the little model spacecraft still hidden away. On 14 October a second booster lifted off the pad. As before, the fourth-stage motor refused to ignite.

Ten days later, on the twenty-fourth, the Soviet Strategic Rocket Forces suffered the worst catastrophe of the space age. A new R-16 booster, the ICBM that would ultimately win out over Korolev's R-9, stood waiting for launch from the Cosmodrome complex at Baikonur. At 6:45 that evening, when the ignition signal was given, the rocket remained silent on the pad. Ignoring every safety rule, Marshal Mitrofan Nedelin, Korolev's military supervisor, led a large party of engineers and technicians out to inspect the vehicle. With men swarming over the pad, chief engineer Mikhail Yangel stepped into a concrete shelter to smoke a cigarette. Suddenly, perhaps because a technician had reconnected an umbilical, the rocket ignited. One survivor recalled the ensuing horror:

At the moment of the explosion, I was about 30 meters from the base of the rocket. A thick stream of fire unexpectedly burst forth, covering everyone around. Part of the military contingent and testers instinctively tried to flee from the danger zone, people ran to the other side of the pad, toward the bunker[,] . . . but on this route was a strip of new-laid tar, which immediately melted. Many got stuck in the hot sticky mass and became victims of the fire. . . . The most terrible fate befell those located on the upper levels of the gantry: the people were wrapped in fire and burst into flames like candles blazing in midair. The temperature at the center of the fire was about 3,000 degrees [5,432 degrees F]. Those who had run away tried while moving to tear off their burning clothing, their coats and overalls. Alas, many did not succeed in doing this.[5]

"Automatic cameras had been triggered along with the engines," dissident scientist Andrei Sakharov reported. "The men on the scaffolding dashed about in the fire and smoke; many jumped off and vanished into the flames. One man momentarily escaped from the fire but got tangled up in the barbed wire surrounding the launch pad. The next moment he too was engulfed in flames."[6] Marshal Nedelin's impatience had cost some one hundred lives, including his own. No longer would Korolev be able to rely on comrade Nosov, who had pushed the button to launch Sputnik 1, or comrade Ostashev, who had played a key role in the R-7's development. Their loss and that of so many of their skilled and dedicated colleagues would be deeply felt during the weeks and months to come.

Failure continued to punctuate the successes of the Vostok program. On 2 December 1960 Korabl Sputnik 3, with the dogs Mushka and Pchelka aboard, was lost during reentry. The next day, Sergei Korolev suffered a heart attack. The physicians treating the chief designer also discovered a kidney problem and suggested extended bed rest. Ignoring their advice, Korolev insisted on returning to work. Four months later, on 23 March 1961, twenty-four-year-old Valentin Bondarenko, the youngest member of the original cosmonaut training-group, was completing a ten-day-long series of tests in a pressurized oxygen chamber. Removing some medical sensors from his body, he swabbed the area with a piece of alcohol-soaked cotton and thoughtlessly tossed the material onto an electric hot plate. The very atmosphere of the chamber burst into flames. His friend, Yuri Gagarin, was by his side when he died of shock eight hours later. Other medical tests and training accidents claimed victims as well. Anatoli Kartashov would be dismissed from the cosmonaut corps in the summer of 1960 as a result of complications from a hemorrhage suffered during a high-speed centrifuge ride.

The tragedy, death, and bitter disappointments were hidden from the West, where most people regarded Maj. Yuri Alexeyevich Gagarin's single orbit of the earth as the latest in a long string of Soviet space triumphs. How much louder would the applause have been had the world known the fate of Marshal Nedelin, Valentin Bondarenko, and so many others? Only those at the center of Soviet power could fully appreciate the raw courage of Yuri Gagarin.

10 • Racing to the Moon

If we do not hitch onto the Moon and quarry our granite there, it won't be the fault of the Yankees.   *Cleveland Abbe, American astronomer and meteorologist, 1866*

A t 10:19 A.M. on 27 March 1968 Soviet air force colonels Vladimir Seryogin and Yuri Gagarin lifted a two-seat MiG-21 UTI trainer aircraft off the runway of the Soviet airfield at Zvyozdniy Gorodok, some twenty-five miles (forty kilometers) northeast of Moscow. The previous November, Soviet authorities had grounded Gagarin, that country's great national hero, to keep him from risking his life unnecessarily. But the personal flight ban had been lifted just two weeks earlier. Relieved of his administrative and ceremonial duties, Gagarin had passed a physical, resumed cosmonaut training, and requalified as a jet-fighter pilot.

The twenty-seventh of March was not a good day for flying. Thick clouds hampered visibility, and a radar malfunction prevented ground personnel from monitoring the altitude or position of nearby aircraft. To make matters worse, air-traffic-control procedures apparently were lax: less than five minutes after Gagaran and Seryogin had taken off a pair of MiG-21s and another MiG-21 were cleared to depart on the same heading. The two jets swept past Gagarin and Seryogin just as their MiG-21 was climbing out of the overcast.

167

Gagarin reported the incident and received permission to return to the air-field. Midway through their turn, however, they encountered the single MiG-21 as it shot up through the clouds, narrowly missing Gagaran and Seryogin's craft. Their own MiG was thrown into a spin and tumbled down through the thick cloud cover. Crash investigators later surmised that the two pilots had succeeded in stopping the spin after five full revolutions. Still disoriented, however, Gagarin and Seryogin never pulled out, diving straight into the ground. Thus perished the first human to have flown into space, a man who, it was widely rumored at the time of his death, was about to lead a crew of cosmonauts in circumnavigating the Moon.

The first human space voyager was the son of a milkmaid and a worker on a collective farm. Yuri Alexeyevich Gagarin was born on 9 March 1934 at Klushino, Smolensk, some 100 miles (160 kilometers) west of Moscow. The circumstances of his growing up would have been familiar to millions of Russians. Nazi invaders reached his village six weeks after he began school, and he spent the next several years living in a dugout. When the Germans withdrew from Klushino, they took his sister and brother with them as hostages. Both eventually made their way home.

Graduating from high school in 1949, Yuri went to work in a factory while pursuing further studies at the Lyubertsy Agricultural Machinery School and the Saratov Industrial Technical School. The combination of a classroom education and practical experience qualified him for a position as a foundry supervisor, but Yuri had other plans. In June 1955 he graduated from the Saratov School with certification as a foundryman-technician and a ground-school diploma from the local flying club. Rather than beginning work in industry, he attended aviation summer camp and was posted to a Red Air Force pilot training course at Orenburg. His fiancée, Valentina Goryacheva, a nursing student, complained that she could never see Yuri when he was marching with his comrades; they buried the shortest men in the center of the formation. His stature remained something of a problem when flight training began. The young pilot habitually flew with a thick cushion beneath his parachute in the pan of the ejection seat.

Gagarin had been out of flight school less than two years when he was ordered to report for the first round of cosmonaut testing. He received notice of his selection as a candidate for space flight on 9 March 1960, his twenty-sixth birthday. By all accounts he quickly became a universal favorite, and the first editor of the *Syringe,* a comic newsletter the cosmonauts produced during their period of medical testing and training. He was just as impressed with his new comrades—"positive heroes," he called them. "I believe that in time our entire country will get to know them," he assured his wife, "and will be justly proud of them."[1]

On 12 April 1961, aboard Vostok 1, Gagarin completed a single orbit of the Earth, remaining aloft for only one hour and forty-eight minutes. The feat had none of the stunning surprise of Sputnik. The world had grown accustomed to Soviet space firsts. The country's three-and-a-half-year string of triumphs

The launch of Vostok 1 on 12 April 1961, which carried Yuri Gagarin into space for an orbital journey lasting just under two hours. As the Russians had hoped, the flight was a propaganda coup. (Courtesy of the Novosti Press Agency)

had exercised a decided impact on public opinion: 41 percent of those surveyed in a contemporary Western European poll regarded the Soviet Union as the most militarily powerful nation on the globe; fewer than 20 percent believed the United States to be superior.

As the Soviets had hoped, the reaction to Gagarin's Earth orbit was particularly strong in African, Asian, and Middle Eastern countries. Kenya's leading newspaper referred to the flight as the most important event since the invention of the wheel. Major journals in Tunisia and Iran compared it to the invention of the printing press and the discovery of America. President Sukarno of Indonesia believed that "the Soviet feat would . . . contribute to the progress and prosperity of mankind as well as to world peace," and Egyptian president Gamal Abdel Nasser congratulated the Soviet Union on its "gigantic scientific capabilities" and predicted that "the launching of man into space will turn upside down not only many scientific views, but also many political and military trends."[2]

On 5 May 1961 American astronaut Alan B. Shepard Jr. followed Gagarin into space. A modified Redstone booster lobbed Shepard's Freedom 7 spacecraft 300 miles (480 kilometers) downrange at a speed of 5,000 miles (8,050 kilometers) per hour. At his maximum altitude of 116 miles (186 kilometers) above the earth, the astronaut was able to change his craft's attitude in yaw, pitch, and roll. In fifteen minutes it was all over. Although the flight was an important success, most people would still have agreed with a Kuala Lumpur, Malaysia, newspaper's statement that "it is evident that the U.S. is losing the space race with Russia."[3]

Just twenty days after Alan Shepard rode Freedom 7 into history, President John Fitzgerald Kennedy addressed a joint session of Congress on urgent national needs. He concluded his talk with an extraordinary challenge. "I believe," he said, "that this nation should commit itself to achieving the goal, before this decade is out, of landing a man on the moon and returning him safely to Earth."[4]

President Dwight D. Eisenhower had been a reluctant spacefarer at best. He had recognized a clear need for military space programs and had approved the first U.S. satellite and manned space initiatives in the interest of science and international public relations. For all that, he had displayed little enthusiasm for a manned space effort, and his administration had offered nothing resembling a long-range, coherent civilian space program.

There was reason to hope that John Kennedy would be more supportive.

On 5 May 1961 Alan B. Shepard became the first American in space. The modified Mercury-Redstone rocket boosted Freedom 7, shown here, to an altitude of 116 miles. Unlike Russian cosmonauts, Shepard was able to adjust the yaw, pitch, and roll of the craft. (Courtesy of NASA)

Three hundred miles down range from Cape Canaveral, Florida, a U.S. Marine helicopter snatches Alan Shepard from the ocean after his 1961 historic fifteen-minute suborbital flight. (Courtesy of NASA)

During his campaign he had emphasized the need to close the supposed "missile gap" and to match the Soviets in space. "To insure peace and freedom," he noted, "we must be first. , . . This is the new age of exploration; space is our great new frontier."[5] Still, the depth of Kennedy's interest was by no means clear. Before taking office the president-elect had asked his science advisers to study the issue of manned space flight and to suggest appropriate policies for the new administration. Like their predecessors of the Eisenhower years, the Kennedy group argued for the development of scientific and applications satellites, but they bluntly advised that "a crash program aimed at placing man into orbit at the earliest possible time cannot be justified solely on scientific or technical grounds."[6]

The attitudes of Kennedy's science advisers underscored a basic tension within the U.S. space program that exists to this day. If the primary goals of the U.S. effort were to achieve purely military objectives, acquire new scientific knowledge, and use the unique orbital vantage point to improve life on

Earth, then it would be unnecessary to send human beings into space. Unmanned satellites and robot lunar and planetary explorers could accomplish those tasks at a fraction of the cost of a manned program.

Kennedy was quick to recognize the larger importance of putting people in space. A young and vigorous man, he came to office heralding a "new frontier," yet the first few months of his administration produced considerable frustration and disappointment. A U.S.-backed invasion of Cuba ended in humiliating defeat, labor unrest at home increased, a crisis was brewing in Southeast Asia, and the new president seemed to come off second best in his first meetings with Premier Nikita Khrushchev and President Charles de Gaulle of France.

John Kennedy saw the space race as a crusade that would rally Americans and restore self-confidence at home while rebuilding national prestige abroad and allowing the United States to stand up to the Soviets without risking nuclear confrontation. The challenge was to choose a spectacular goal in space that the United States could be sure of attaining before the Soviet Union. Secretary of Defense Robert S. McNamara believed that the Russian lead in space was so great that the United States could not be certain of winning anything short of a race to Mars. Vice President Lyndon B. Johnson, however, assured the president that a journey to the planets would not be necessary: NASA could beat the Soviets to the Moon.

Such a program would come with an enormous price tag: initial estimates were as high as $24 billion. Uncertain as to how the nation would react to such an expensive project, Kennedy presented the Apollo Moon-shot program to Congress and to the American people as a tempting challenge. The sheer excitement of the idea instantly captured the public's imagination. Congress voted the funds and the taxpayers applauded.

Scattered voices of dissent eventually were raised. Civil-rights leaders and social activists questioned the value of a trip to the Moon at a time when the nation continued to face serious domestic problems, and many scientists found it difficult to work up much enthusiasm for a program in which the journey itself was more important than the knowledge that it could generate. Yet most Americans, caught up in an exciting drama complete with a little band of heroic patriots who faced cosmic dangers, were not listening to the skeptics.

Perhaps more than any other individual, James Edwin Webb deserves ultimate credit for the success of the U.S. space effort. Appointed in 1961 to re-

President John F. Kennedy recognized the political and symbolic value of the space program. Here the president speaks with the prime minister of Nigeria via the Syncom communications satellite on 23 August 1963. (Courtesy of NASA)

place T. Keith Glennan as NASA's administrator, Webb was a professional manager with long experience in business and government. As a former marine pilot and a member of the board of directors of McDonnell Aircraft, he was no stranger to aerospace. President Kennedy, the Congress, and the American people had decided to go to the Moon, and James Webb was just the man to get the job done. A genuine technocrat, Webb believed wholeheartedly that the power of a modern nation depended on its ability to manage large-scale technical enterprises. The great question of the Cold War, he believed, was whether the United States, a capitalist republic, could organize a vast and complex technological enterprise as successfully as a highly centralized state like the Soviet Union, in which critical decisions were made by a handful of men.

A master politician as well as an excellent manager and administrator, Webb knew precisely how to win support for his program, as Robert Gilruth, the head of Project Mercury, learned during the planning for the Manned Spacecraft Center in Houston. Facing a period of enormous growth, NASA required a new facility to house the manned space effort. When Webb announced that such a center would be constructed for that purpose near Houston, Gilruth suggested that it might be easier and less expensive to expand the existing Langley Research Center in Hampton, Virginia. "Bob," Webb replied, "what the hell has Senator Harry Byrd [of Virginia] ever done for you or NASA?" Houston, he pointed out, was in the congressional district of Representative Albert Thomas, chairman of the House Appropriations Committee, which controlled NASA's budget. He scarcely need have added that Texas also was the home state of Vice President Lyndon Johnson, one of the space program's most enthusiastic supporters.[7]

Webb and his headquarters staff ultimately were responsible for administering every phase of the Apollo program, they set the schedules, established the budgets, and monitored performance. Webb had every confidence in the scientists and engineers working at the various NASA centers. Nevertheless, he maintained an independent Office of Reliability and Quality Assurance, whose job it was to spot problems and report them directly to him.

Deciding to go to the Moon was one thing; deciding *how* such a journey might be accomplished was something else again. Clearly, a very large rocket would be required for the trip. And since the mid-1950s, building just such a rocket had been the dream of Wernher von Braun and the staff of what had become NASA's George C. Marshall Space Flight Center in Huntsville, Alabama. As early as the spring of 1957, inspired by a Department of Defense specification for a launch vehicle capable of orbiting heavy military satellites, von Braun's team had begun design studies for a "Super Jupiter" with a first-stage thrust of 1.5 million pounds (6,700,000 newtons).

Engineers at the Rocketdyne Division of North American Aviation designed two engines that could power such a giant's main stage: the 360,000- to 380,000-pound-thrust (1,601,423- to 1,690,391-newton) E-1 and the even more powerful 1,600,000-pound-thrust (7,117,437-newton) F-1. These engines would not be ready for several years, however. In the interim von Braun and his colleagues planned a Juno 5 that would be powered by a cluster of eight H-1s, increased-thrust versions of the Jupiter engine. The project received approval from DoD's Advanced Research Projects Agency in August 1958, at which time von Braun renamed the proposed new rocket Saturn.

The question of the Saturn's upper stages was also under study. When NASA took over the Army Ballistic Missile Agency in the spring of 1960, von Braun's personnel were considering employing Atlas or Titan boosters on top the large cluster of engines and tanks that would make up the rocket's first stage. Abe Silverstein, a propulsion expert serving as NASA's first director of space-flight development, proposed the idea of a revolutionary new engine that burned liquid hydrogen and liquid oxygen.

Rocket pioneers Konstantin Tsiolkovsky, Robert Goddard, and Hermann Oberth had recognized that a combination of liquid hydrogen and liquid oxygen was the ideal propellant, producing a specific impulse—the standard measurement of rocket performance—some 40 percent higher than that attained by a mixture of liquid oxygen and kerosene. But handling, housing, moving, and burning liquid hydrogen, at temperatures below -423 degrees F (-253 degrees C), would create daunting problems for engineers. Recalling the difficulties that propulsion specialist Walter Thiel had experienced with an experimental hydrogen engine in 1937, von Braun had no desire for the members of his team to take on such a project. Nevertheless, a NASA committee headed by Silverstein finally persuaded von Braun to work on a liquid-hydrogen upper stage, a task that was rendered less intimidating by NASA's hiring of a contractor with considerable experience in this new fuel technology.

In 1956 Pratt and Whitney had developed the 304-2, a super-secret, hydrogen-fueled jet engine for Project Suntan, which was to result in a revolutionary new high-performance reconnaissance aircraft. When the project collapsed two years later, the company decided to put its hard-won experience with the volatile propellant to work in the design of a rocket engine. The resulting RL-10 engine was used to propel a new upper stage called Centaur. Mated to an Atlas, Centaur was capable of orbiting relatively heavy payloads or sending lighter weight instrument packages into deep space. From the outset, Saturn was envisioned as a family of rockets. Early vehicles in the series would feature a first stage consisting of a cluster of H-1s and a second stage powered by RL-10s. The pinnacle of the Saturn group, a behemoth called Nova, would be boosted off the pad by a first-stage cluster of from eight to twelve F-1 engines capable of generating as much as 19,200,000 pounds of thrust (85,409,252 newtons).

Ultimately, of course, the parameters of the mission would set the requirements for the launch vehicle. There was more than one way to fly to the Moon. The 600-foot (183-meter) Nova "dream rocket" could boost an astronaut crew directly to the Moon in a single spacecraft so large and powerful that it would be able to land on the lunar surface and then take off and fly directly back to Earth. Wernher von Braun and virtually all of the other space dreamers had always envisioned just such a "direct-ascent" mode for the journey.

There was, however, a problem. Von Braun, the best judge of the matter, had serious doubts as to the ability of his team and their contractors to design, build, and test the Nova in time to meet President Kennedy's goal. Instead the veteran rocket engineer suggested an Earth-orbit-rendezvous (EOR) scheme that would employ two smaller Saturn rockets: one launcher would carry most of the propellant into orbit; the other would boost the crew aloft. The two halves of the vehicle would dock in Earth orbit and proceed together on their journey to the Moon.

Robert Gilruth and his Space Task Group members, now relocated to the new Manned Spacecraft Center in Houston, resisted the proposal. "I feel that it is highly desirable to develop a launch vehicle with sufficient performance and reliability to carry out the lunar landing mission using the direct approach," he stated, adding that he saw von Braun's plan as a "crutch to avoid the difficulty of developing a reliable Nova class launch vehicle."[8]

While the battle lines between Houston and Huntsville were still forming, John C. Houbolt, a NASA mathematician based at Langley, offered a third approach to the problem. Von Braun had suggested using two rockets to launch the complete lunar craft. Houbolt argued that the spacecraft's weight and size could be reduced sufficiently to allow it to be launched by a single Saturn. Composed of two distinct spacecraft—the command-and-service module (CSM) and the lunar module (LM)—Houbolt's proposal was dubbed the lunar-orbit rendezvous (LOR). The entire assembly would be launched together and would travel to the Moon with the crew riding primarily in the CSM. Once they were safely established in lunar orbit, the astronauts would enter the LM, disengage from the orbiting CSM, and descend to the Moon's surface. Their work completed, the crew would blast back into orbit in the top half of the LM, leaving the bottom half on the lunar surface. The

Christopher Kraft *(left),* director of flight operations; astronaut Gordon Cooper
(center); and Robert Gilruth, director of the Manned Spacecraft Center, Houston,
celebrate a Gemini triumph in the best NASA style. (Courtesy of NASA)

astronauts then would rendezvous with the waiting CSM, reenter it, discard
the LM's top half, and return home in the CSM.

Lunar-orbit rendezvous would enable NASA engineers to conduct the
mission with much less propellant—and, therefore, much less weight—than
would be needed for either a direct ascent or an Earth-orbit rendezvous,
as only part of the LOR spacecraft would land on the lunar surface or be
boosted back into space. The greatest disadvantage was obvious: rendezvous
and docking, the mission's most complex elements, would be undertaken
240,000 miles (386,000 kilometers) from Earth. If something went wrong in
lunar orbit, there would be no hope of rescue or assistance.

Houbolt explained his LOR plan in a letter to NASA's associate administrator, Robert Seamans, who saw it as a way to save time and also resolve
some of the difficulties between Houston and Huntsville. By the spring of
1962 Gilruth had come out in favor of LOR, and Wernher von Braun, recognizing the strength of the forces arrayed against him and the need to foster cooperation for the greater good of the program, held out only a few weeks
longer.

Meanwhile, in the Soviet Union, Sergei Korolev also was giving considerable thought to the Moon. It is not clear whether the Soviet decision to attempt a lunar mission preceded President Kennedy's call to arms or was inspired by it. In any case the Moon was the obvious goal of the space race once the contest had been joined. For Korolev and his Kremlin superiors, there was little choice but to forge ahead. By the end of 1961 Korolev had developed plans for two new boosters that were capable of either lifting heavy payloads into Earth orbit or sending a minimum manned expedition to the Moon. The N-1—to be designed, built, and tested between 1962 and 1965—would be followed into service in 1970 by the larger and more powerful N-2. As with von Braun in the United States, however, Korolev faced political problems and interpersonal feuds that had marred Soviet rocketry since the Stalinist era.

Korolev's old rival, Vladimir Chelomei, and his design bureau proposed building a heavy-lift booster to be known as the UR-700, whereas Mikhail Yangel promoted the R-56 design. In July 1962 a state commission headed by academician Mstislav V. Keldysh recommended that Korolev be allowed to press ahead with the development of his N-2 and that Chelomei continue work on a smaller design. The Soviets may have been hedging their bets as they encouraged this competition between rival engineering teams. Korolev's N-2 could serve as the basis for a minimum-landing expedition to the Moon, and the less powerful UR-500 could at least propel cosmonauts on a circumlunar journey.

Korolev was also struggling to hold his own team together. His propulsion specialist, Valentin Glushko, favored the use of nitrogen tetroxide and unsymmetrical dimethyl hydrazine (UDMH) as propellants for the upper stages of the N-1 and N-2 vehicles; Korolev preferred liquid oxygen and liquid hydrogen. The dispute led to the end of the most productive and difficult working relationship in the history of Soviet rocketry. The most knowledgeable Western commentator on the Soviet lunar program dates the ultimate failure of that effort to the collapse of this thirty-year relationship.

Glushko was reassigned to Chelomei's team, and it was decided that the main-stage engines of the UR-500 would feed on his preferred propellants. Korolev had not only lost the services of the USSR's finest engine designer but also earned the enmity of a man with a powerful voice in the rocket community. The hero of both the Sputnik and Gagarin flights, Korolev now was to receive rocket engines from N. D. Kuztenzov's aircraft-design bureau. With no prior experience in rocket design, the new team was reluctant to work with liquid hydrogen. Korolev's N-booster, like its R-7 predecessor, would burn the standard liquid oxygen and kerosene.

By the end of 1964 Korolev and his staff were struggling to work with their new colleagues at Kuztenzov while putting the final touches on their plan for a mission to the Moon. Unknowingly, Soviet engineers had set off down a path blazed by their U.S. rivals. Discarding an initial plan to launch their spacecraft aboard two boosters with a rendezvous in Earth orbit, they opted for a multipart spacecraft sent aloft aboard a single booster. Like the Americans, they would attempt a lunar-orbit rendezvous.

While the two most powerful nations on Earth laid plans for a trip to the Moon, their citizens continued to celebrate the astronauts and cosmonauts who ventured into space aboard single-seat Mercury and Vostok spacecraft. Both the U.S. and Soviet programs had relatively simple goals: to test equipment and procedures, to explore the physical and psychological reactions of human beings to space flight, and to demonstrate the power and prestige of their respective nations.

On 21 July 1961 Virgil I. "Gus" Grissom blasted off the launch pad aboard the Liberty Bell 7, a Mercury spacecraft similar to that in which Alan Shepard had flown. At the conclusion of its suborbital flight, the spacecraft was bobbing safely in the Atlantic when its emergency escape hatch blew off, forcing

Table 1

Vostok and Project Mercury Manned Missions

Mission	Dates	Crew	Duration (Days:Hours:Minutes)
Vostok 1	12 April 1961	Yuri Gagarin	0:1:48
Freedom 7	5 May 1961	Alan Shepard	0:0:15
Liberty Bell 7	21 July 1961	Virgil Grissom	0:0:15
Vostok 2	6–7 August 1961	Gherman Titov	1:1:11
Friendship 7	20 February 1962	John Glenn	0:4:55
Aurora 7	24 May 1962	M. Scott Carpenter	0:4:56
Vostok 3	11–15 August 1962	Andrian Nikolayev	3:22:9
Vostok 4	12–15 August 1962	Pavel Popovich	2:22:44
Sigma 7	3 October 1962	Walter Schirra	0:9:13
Faith 7	15–16 May 1963	L. Gordon Cooper	1:10:20
Vostok 5	14–19 June 1963	Valery Bykovsky	4:22:56
Vostok 6	16–19 June 1963	Valentina Tereshkova	2:22:44

Astronaut John Glenn rides down Broadway as thousands of New Yorkers celebrate his successful Earth orbit of 20 February 1962. (Courtesy of NASA)

Grissom to exit hastily as water poured into the cockpit. The astronaut was
rescued, drenched and exhausted, but his water-filled spacecraft, too heavy
for the Navy helicopter to lift, was lost.

An Atlas rocket boosted John Glenn off the Cape Canaveral launch pad at
9:47 on the morning of 20 February 1962. Everything seemed to be going
well on the first U.S. orbital mission until telemetry signals from the Friendship 7 spacecraft suggested to ground controllers that its all-important heat
shield had come loose. The nation held its collective breath as Glenn dropped
back toward Earth and disappeared into the normal radio blackout accompa-
nying reentry. The heat shield held, and astronaut John Hershel Glenn
emerged as the hero of the U.S. space program.

Six of the country's original seven astronauts flew into space between 5
May 1961 and 15 May 1963. The last, the late Donald K. "Deke" Slayton,
grounded before he could fly on a Mercury mission by a minor heart irregu-
larity, joined this exclusive club in 1975, when he served as a pilot on the joint
U.S.-Soviet Apollo-Soyuz Test Project (ASTP) flight.

Project Mercury offered moments of high drama. Everything seemed to go wrong for Malcolm Scott Carpenter, who followed Glenn into space. The experiments he was to perform aboard his craft, Aurora 7, malfunctioned, and his body temperature rose to 102 degrees F (38.8 degrees C). "I did not believe that I was actually running a fever," Carpenter recalled, "but I did begin to notice one of the first symptoms. . . . I was having trouble finding the right words with which to express myself in my reports to the ground."[9]

Distracted, Carpenter fired his thrusters time and again in an attempt to control the attitude of his spacecraft, thus reducing his supply of propellant to a dangerous level and forcing ground controllers to consider ending the flight prematurely. When the time to return to Earth finally did arrive, the retro-rockets failed to fire automatically, and Carpenter had to trigger them manually, three seconds late, with the spacecraft a full twenty-five degrees out of proper alignment. Carpenter splashed down some 250 miles (400 kilometers) from the projected landing area and bobbed around in a life raft until his discovery by a very worried search party. "It seemed a pity that I was having to spend so much time worrying about a man-made object," the astronaut commented about his space-flight experience, "when God's own creations, just outside the window, were much more mysterious and challenging."[10]

In Aurora 7's aftermath Wally Schirra was determined to conduct a text-book mission, and he did just that, landing only five miles (eight kilometers) from the splashdown target with 75 percent of his thruster propellant still aboard. Soon thereafter Gordon Cooper, the youngest astronaut, made the last—and the longest—flight in the Mercury program. Thirty years later he still holds the U.S. record for solo flight time in Earth orbit and is the only astronaut known to have fallen asleep in his spacecraft during countdown.

Soviet cosmonauts ventured aloft less frequently than their American counterparts during this period, but they invariably remained in space longer. Soviet space planners did their best to steal headlines from NASA. Twice they launched Vostok spacecraft only a day apart. The vehicles could not maneuver to rendezvous in orbit, but they did fly in tandem only three to four miles (five to six kilometers) apart. Although the missions of Vostoks 3, 4, 5, and 6 did not represent any great step forward, the fact that two spacecraft had been launched into precise orbits on two consecutive days did point up the reliability of the Soviet program, and it looked good in the papers.

The pilot of Vostok 6, the last flight in the series, was Valentina V. Tereshkova, the first woman to fly into space. The flight's primary goal was clearly publicity: Tereshkova, one of four women who had entered cosmonaut training in March 1962, was a sport parachutist, not an experienced jet pilot. However, in view of the fact that most of the Vostok cosmonauts were little more than well-trained passengers aboard automated craft, the absence of significant flying experience was no great handicap.

The Vostok flights were impressive, but Americans obviously had recovered from their crisis of confidence. Project Mercury had more than fulfilled its mission, offering repeated public demonstrations of NASA's ability to send human beings into orbit and return them safely to Earth. From start to finish, it had cost the U.S. taxpayer some $400 million to send six of their countrymen into space. It had been money well spent. As veteran space voyager Michael Collins pointed out, "The word 'astronaut' had become a totally understood and accepted term in the American vocabulary, and millions of kids aspired to become one."[11]

11 • The Giant Leap
Gemini

n the spring of 1962 thirty-one-year-old Michael Collins, a captain in the U.S. Air Force, was outranked by virtually every male member of his family, beginning with his uncle, U.S. Army chief of staff J. Lawton Collins. Collins's father had retired as a two-star army general. His brother, another uncle, and his cousin held the ranks of Army colonel, brigadier general, and major, respectively.

After graduating from the U.S. Military Academy at West Point, New York, Collins chose a new direction for the family military tradition by accepting a commission in the U.S. Air Force. He earned his wings in 1953 and spent the next seven years paying his dues in the time-honored fashion of all junior officers. Collins's ticket to the inner sanctum—a posting to the Experimental Flight Test Pilot School at Edwards Air Force Base, California—came in 1960.

Most of the traditions honored at Edwards were less than fifteen years old, but they ran very deep. This was where they flew higher and faster than anywhere else in the world. The first U.S. jets had been tested at Edwards, and experimental aircraft, from the faster-than-sound Bell X-1 to the North American X-15 that would fly to the edge of space, had made contrails in the clear, blue skies overhead. The men who flew these planes were legends. Some, like Chuck Yeager, Scott Crossfield, and Pete Everest, were still around. Others had died here. The dusty streets that crisscrossed the base bore their names.

Captain Glen W. Edwards, for whom the base was named, had lost his life in the 1948 crash of a Northrop YB-49 Flying Wing.

The members of class 60-C entered the Experimental Flight School beneath a sign that read, "Through these portals pass the world's finest pilots." Mike Collins thought that had a nice ring to it, and so, presumably, did the thirteen other "mostly hyperthyroid, superachieving first sons of superachievers" who made up his class. To this day, Collins recalls in his book *Carrying the Fire*, "I am impressed by this group."[1] Over the next ten years one of them, James Irwin, would walk on the Moon, and two others, Collins and Frank Borman, as well as one of their instructors, Thomas Stafford, would orbit it.

But all of that lay in the future. In the late summer of 1960 the veteran pilots of Edwards's famous Fighter Ops unit were not impressed by what they had heard about Project Mercury. "Man, they were here to fly, not to be locked up in a can and shot around the world like ammunition," Collins explained. "They were master craftsmen, artists, they were Jonathan Livingston Seagulls—they flew, in smooth control, in command."[2] The newcomers were not so sure. President Kennedy's 1961 challenge of a flight to the Moon cast the venture in a very different light. What difference did it make that the astronauts were blasted into space aboard a craft without wings and returned to Earth dangling beneath a parachute? What the veterans saw as an intolerable affront to their dignity as intrepid aviators, the younger men regarded as a small price to pay to stand on the surface of another world.

In the spring of 1962 word spread through the test-pilot fraternity that NASA was recruiting a second group of astronauts. Mike Collins submitted his application "before the ink was dry on the announcement." Frank Borman and Tom Stafford were not far behind. Determined to put its best foot forward, the air force established an internal screening process. Applicants who passed an initial interview were called to Washington for a pep talk from Gen. Curtis LeMay, the air force chief of staff, and for advice on everything from elocution to the importance of pointing the thumbs to the rear when standing with hands on hips. Thus inspired, the young air force officers went on to Brooks Air Force Base, Texas, where they were subjected to the same round of physical and psychological tests pioneered by the Mercury Seven. Borman and Stafford were among nine selected. Undaunted, Collins repeated the process one year later and, along with thirteen other rookie astronauts, reported for duty in January 1964 at the new NASA Manned Spacecraft Center in Houston.

As early as 1960 Jim Chamberlain, a Canadian engineer in Bob Gilruth's Space Task Group, had begun to give serious thought to a new generation of manned spacecraft, originally called Mercury Mark II. Chamberlain envisioned a two-seat craft capable of remaining in orbit for as long as two weeks and ultimately, perhaps, able to fly around the Moon. The decision to land on the Moon led STG planners to redefine the role of the Mark II. The ability to manage a rendezvous between two spacecraft would be key to the success of a lunar mission. While the enormous rockets and complex spacecraft required for the Apollo mission were being designed and built, the new two-man craft would maintain a U.S. presence in space and test the on-board computers, radar systems, and docking mechanisms that would enable two small craft to come together in the depths of space. James Webb's new associate administrator, Robert Seamans, approved the plan and reapportioned $75 million from the 1962 budget appropriation to set things in motion. The selection of the manufacturers of the Mercury spacecraft, McDonnell Aircraft of St. Louis, to construct the new vehicle also would speed the process.

Gemini, as the craft would be known, bore an external resemblance to Mercury but was actually quite different. All of the required equipment had been stuffed inside the tiny Mercury vehicle. Gemini would feature an adapter section or service module behind the spacecraft designed to carry a fourteen-day supply of oxygen, propellant for the thrusters, and electrical generating equipment. The entire module would be jettisoned prior to reentry.

Designed to be launched by an air force Titan 2 rocket, the Gemini spacecraft would feature new electronic gear that would enable it to perform its primary mission: to locate and dock with an unmanned Agena vehicle orbited by a separate Atlas rocket. It also would test other systems required for long-duration missions. The Mercury spacecraft's batteries, for example, were too heavy and short-lived for a flight to the Moon, whereas Gemini's electrical needs would be met by special fuel cells.

Because astronauts would have to leave the protection of their ship to walk on the Moon, Gemini would blaze another trail. Whereas the Mercury spacesuit had been essentially an item of emergency equipment—unnecessary unless something went wrong—the Gemini suit was essential. During a Gemini mission, the entire spacecraft would be depressurized to allow one of the astronauts to open the large overhead hatch and float free in space at the end of a 50-foot (15-meter) umbilical. During this extravehicular activity (EVA), spacesuits would be the only items protecting the astronauts from the harsh environment of space.

The Gemini spacecraft was not designed for comfort. The cockpit was so tiny that any thought of moving about or changing clothes was out of the question. Some crews would spend as long as two weeks with less space per man than a phone booth, without washing, stretching, or shaving. It was not a pleasant prospect.

By the end of 1963 administrator James Webb and his colleagues began to wonder when the first two astronauts would shoehorn themselves into the tiny Gemini cockpit for a flight. The program had fallen far behind schedule. Problems with the Titan 2 booster caused some concern, but the real difficulties lay in the radically new system devised for landing the spacecraft. Because of their purely ballistic flight paths, the slightest error in retrofire brought Mercury spaceships back to Earth as many as 250 miles (403 kilometers) from their planned landing point. In an effort to remedy that situation Gemini was designed to develop some lift from air passing over the heat shield, lift that could be translated into some measure of control during reentry. In addition, Jim Chamberlain's team planned to furnish Gemini with landing gear and a flexible wing that could be deployed during descent Rather than splashing into the ocean beneath a parachute, the crew would glide back to a wheels-down landing on a runway. Unfortunately, the engineers had not been able to get the system to work.

Enough was enough. NASA headquarters in Washington reassigned Chamberlain and placed Charles Mathews, another STG veteran, in command of the Gemini program, with instructions to get things back on track. Mathews dropped the parasail in favor of a traditional splashdown and took additional steps to simplify the program and meet the schedule. On 8 April 1964 the first unmanned Gemini test vehicle finally roared aloft atop a Titan 2. The first manned flight was still a year away, but progress was apparent.

Sergei Korolev's plan for a lunar mission was both less detailed and less ambitious than that of his U.S. rivals. With work underway on a new generation of launch vehicles, he turned his attention to spacecraft design. And although NASA and the U.S. military knew very little about the Soviet space program, many details of the U.S. program were available in public literature, and Korolev was able to stay up to date with his competitors.

In 1960, prior to the Earth-orbit-rendezvous versus lunar-orbit-rendezvous debate, NASA had called for preliminary design proposals for a three-part spacecraft capable of flying to the Moon. General Electric suggested a vehicle that would be assembled in Earth orbit. It included a mission module, in which the crew would be launched and in which they would spend most of the journey; a descent module, complete with a heat shield, in which the crew would return to Earth; and a propulsion module to propel the entire assembly to the Moon. NASA did not fund the proposal, but GE's idea sparked considerable discussion, and Korolev seems to have been impressed as well. From 1962 to 1964 the Soviet engineer was studying a three-part lunar vehicle that also would be assembled in Earth orbit. Presumably a fourth element, a lunar lander, would have been added to the package. The entire assembly was to be known as Soyuz ("union").

It was, of course, a complex, time-consuming, and extraordinarily expensive plan. Like the Americans, Korolev eventually chose a simpler scheme based on a lunar-orbit rendezvous (LOR), although the basic Soyuz crew module survived as the key element of that plan. Work continued on the design of the new spacecraft, which was scheduled for its first Earth-orbital launch in 1966.

Ultimately, political commitment to a Moon race was much less clear in the Soviet Union than in the United States. The Kennedy administration took a rational and relatively long-term view, carefully selecting the Moon as a goal that the U.S. could achieve more quickly than could the USSR. NASA officials simply moved forward, step-by-step, confident that they would keep up with the Soviets at every stage and eventually pass them before the time had come to land on the Moon. Nikita Khrushchev's goal was much simpler: to beat the Americans at every opportunity. By 1963 Soviet space spectaculars were an indispensable element of foreign policy. The Cuban missile crisis of 1962 placed the Soviet premier in a particularly difficult position. The decision to withdraw Yangel's R-12 missiles from Cuba represented a retreat from American power. Faced with a strong U.S. challenge abroad and opposition to his policies at home, any further impression that the United States was pulling ahead of the USSR in space would be disastrous.

By the mid-1960s, however, the illusion of Soviet leadership in space became even more difficult to maintain. The early Soviet successes had set an

American technological juggernaut in motion. Korolev and his team were running as fast as they could, and they were falling behind. Moreover, the U.S. space program was conducted in full view of the world. Although this occasionally resulted in a very public failure, it also underscored the drama of the enterprise and generated global interest and excitement.

"Just imagine," Korolev remarked to a colleague during the first difficult months of the American Mercury program, "that there should be a report in one of our papers that the flight of Vostok 2 was due to take place today, but that because of some technical fault we were putting it off until tomorrow." When his friend chuckled at the thought, Korolev became angry. "What the hell are you laughing at? You ought to be crying!"[3]

Now, although it might be slow in coming, the United States' Gemini program loomed on the horizon, and Premier Khrushchev, unwilling to wait for the 1966 appearance of Soyuz, demanded an interim program. As Korolev analyzed Gemini he saw that it involved two astronauts operating a maneuverable spacecraft on missions that would include rendezvous, docking, and extravehicular activity. He could not match that, nor was there time to develop a new interim vehicle like Gemini. By radically modifying the Vostok module, however, he might still be able to produce a surprise or two for the Americans.

There was no way to provide Vostok with the capability to maneuver, change orbits, or dock. Even stuffing an extra crewman into the vehicle would be a difficult proposition involving considerable risk, as Vostok capsules landed at a very high velocity with considerable impact. For that reason, all of the Vostok cosmonauts had ejected from their spacecraft before they landed. The Soviets had yet to admit this fact, fearing that cosmonauts would be disqualified from world flight records. The Federation Aeronautique International required a valid record-holder to land aboard his or her craft. There was no way to fit a second ejection seat into the Vostok. Besides, with the ejection seat removed, Soviet engineers discovered that they could fit three cosmonauts into the capsule—as long as they were not wearing spacesuits. Korolev decided to take the chance, developing a small, solid-rocket motor that would deploy beneath the main parachute and would fire to reduce the spacecraft's velocity just before impact. The risk would be enormous, but the opportunity to fly three cosmonauts before NASA orbited its first two-man Gemini was politically irresistible.

Cosmos 47, the unmanned test-version of the new spacecraft, blasted off from Tyuratam on 6 October 1964 and was safely recovered the following day. Six days later, on 12 October, Boris Yegorov, Vladimir Komarov, and Konstantin Feoktistov, one of Korolev's design engineers, roared aloft aboard the first Voskhod ("sunrise"). Sergei Korolev must have breathed an enormous sigh of relief when the three men returned safely to Earth on the thirteenth.

If rumors circulating at the time are to be believed, the Voskhod 1 flight was going so well that Komarov requested an extension of their time in orbit. Korolev ordered them down with a quote from *Hamlet:* "There are more things in Heaven and Earth, Horatio, than are dreamt of in your philosophy." The three cosmonauts returned to discover that great events were indeed underway in the Kremlin. The next day, Khrushchev was removed from power.

Korolev hoped the new leaders, Communist Party chairman Leonid Brezhnev and Premier Aleksey Kosygin, would relieve some of the pressure on his program. He argued that the emphasis on the public relations value of space flight was shortsighted and suggested the development of a plan for slow, steady growth that would include long-duration flights in Earth orbit, a space station, and ultimately, a trip to the Moon. The Soyuz spacecraft was the key to such a plan and, as such, ought to be given the highest priority.

Brezhnev and Kosygin were apparently sympathetic, but they could ill afford to allow the Gemini to boost the United States ahead of the Soviet Union so soon after they had come to power. Korolev returned to work, determined to produce one more great moment in space history for his new superiors. He would find a way to allow a cosmonaut to leave Voskhod and float free in space. Arranging a spacewalk would be difficult. Unlike the Gemini spacecraft, which would be filled with circuit boards and advanced electronic packages, Voskhod's instruments still employed vacuum-tube technology. Exposing a cabin full of hot tubes to the frigid vacuum of space so that a cosmonaut could exit for a spacewalk would be an invitation to disaster.

The mission would require yet another redesign of the Vostok. This time spacesuits obviously would be required, so only two cosmonauts could fit in the spacecraft. In order to avoid depressurizing the cabin an inflatable airlock would be attached to the outside of the spacecraft over the main hatch. Once the cosmonauts were safely in orbit, they would inflate the device and open the hatch. One crew member would crawl up into the structure then turn and close the spacecraft hatch. He could then dump the air, open a hatch on the airlock, and float out into space without exposing the cabin to loss of pressure or temperature. Experience would prove that the design of Voskhod 2 was every bit as dangerous as it sounded.

Surviving members of the Soviet team recall that Korolev and Leonid Voskresensky, his chief spacecraft designer, worked themselves to exhaustion during the winter of 1964–65. With the first Gemini launch scheduled for the spring of 1965, their time was short. Korolev's doctors repeatedly expressed concern over his failing health. One friend reported that the man who had provided the Soviet Union with some of its greatest triumphs would come home at night too tired to climb the stairs to his apartment. In January 1965 Voskresensky died of a heart attack. Korolev gave the eulogy at his colleague's funeral, then pushed on alone.

As with Voskhod 1, the first test of the new spacecraft would be an unmanned flight disguised as a satellite launch. Sent aloft on 22 February 1965, Cosmos 57 broke up during its second Earth orbit. Over the next week, Western radar stations tracked the reentry and destruction of some 180 bits and pieces of the craft.

The fact that Alexei Leonov and Pavel Belyayev were launched aboard Voskhod 2 on 18 March 1965, less than a month later, indicates both the desperate desire of Soviet leaders to stay one step ahead of the Americans and the utter exhaustion of the Korolev team. As soon as they were in orbit, Belyayev began to assist Leonov into his life-support pack and to attach the umbilical that carried his primary air supply and communications cable. The exit through the airlock went smoothly. Leonov floated into space, but his umbilical was twisted. For ten minutes he fought to overcome the torque of the life-support tether, his gyrations recorded by a low-resolution television camera that he had attached to the top of the airlock. When he attempted to reenter the airlock, he found that his suit had ballooned to such an extent that he could no longer bend his legs to squeeze into the opening. His heart rate and body temperature were rising, and the situation grew desperate. Belyayev, who could not see what was going on, heard Leonov's every word. "I can't. . . . No again, I can't get in. . . . I can't." But there was nothing he, or anyone else, could do to help. Finally, Leonov was able to lower the pressure in his suit and work himself back aboard.[4]

There was more excitement to come for the crew of Voskhod 2. The main retrorocket refused to fire when they attempted to drop out of orbit on 19 March. Approaching the next available landing site, they tried an emergency backup rocket motor. The pair landed safely in a snow-covered forest in the western Ural Mountains, roughly 2,000 miles (3,220 kilometers) from where the recovery crews were expecting them, and spent the night shivering in the spacecraft while a pack of wolves circled in the darkness. The cavalry, in the form of ski troops, arrived the next morning.

A rumored Voskhod 3 mission did not come to pass. Perhaps the barely averted multiple disasters of Voskhod 2 frightened mission planners. The last flight in the series, Cosmos 110, sent two dogs into orbit for twenty-one days. For the next two years, the American astronauts would have Earth orbit all to themselves.

But there was to be no rest for Sergei Korolev. If future Soviet manned missions had to wait for Soyuz, then that program would have to move forward with all possible speed. There was also much to be done to upgrade the SL-4, the latest version of the R-7, for Soyuz test flights. There was even some talk of attempting to send a single cosmonaut looping around the Moon in October 1967 to celebrate the fiftieth anniversary of the Bolshevik Revolution. Moreover, other design bureaus were hard at work on rocket programs that would be crucial to the lunar effort.

Vladimir Chelomei, now recognized as a Soviet military rocket designer, tasted genuine success with the launch of his SL-9, the simplest variant of the long-awaited Proton, in the summer of 1965. It became increasingly apparent that although Korolev would be responsible for future manned spacecraft, Chelomei and others would build the next generation of rockets that would carry them into space.

Sergei Pavlovich Korolev's long, difficult journey came to an end in January 1966. Diagnosed as suffering from a bowel obstruction, he was admitted to a special Politburo hospital in the Moscow suburb of Kuntsevo. Dr. Boris Petrovskiy, the Soviet minister of health, performed exploratory surgery. Discovering a large malignant tumor during the course of the procedure, Petrovskiy elected to remove it, a task for which his team was not fully prepared. Korolev died on the operating table. Two days later, on 16 January, Pravda printed his obituary. In announcing his death from cancer complicated by emphysema and arteriosclerosis, the newspaper also made the first public recognition of his existence and his achievement. The man whose life mirrored the tragedy of his nation's history was accorded full state honors, including the burial of his ashes in the Kremlin wall. He was fifty-nine years old.

Books and articles extolling Korolev's genius suddenly flooded the market. The Soviet film industry produced a cinematic epic based on his life. He was, of course, portrayed as a socialist hero, a man who knew the great Tsiolkovsky and shared his dream, who built weapons for the defense of the Motherland, produced the incredible miracle of Sputnik, and served as a father figure for the cosmonauts. In life the state had insisted on his anonymity. His triumphs had been portrayed as victories of the Soviet people and their farsighted leaders. In death, the Soviet government transformed him into a legend: a plaster Soviet saint drained of blood or life. The best part of the man—the indomitable will and strength of character that had enabled him to survive Stalinist purges, life in a prison camp, and the enormous pressures of work in a brutal Cold War bureaucracy—remained closely guarded state secrets.

In the United States 1965 was the year of Gemini. On 23 March veteran astronaut Gus Grissom and rookie John Young became the first men to ride a Titan 2 into space. For the first time, a manned spacecraft changed its orbital plane by firing its thrusters. After orbiting the earth three times Gemini 3 landed fifty miles (eighty-one kilometers) from its aiming point. Gemini proved to be less maneuverable during reentry than its designers had hoped, but a more serious problem occurred at the flight's conclusion. The spacecraft's parachute harness was designed to shift the vehicle from a vertical to a horizontal position for splashdown. It worked, but the sudden change in attitude was so violent that Grissom cracked his faceplate against the instrument panel.

Unfortunately, the presence of a sandwich aboard the spacecraft would overshadow the real technical achievements of the flight. The crew had been assigned to test some food-packaging systems designed for long-duration missions. Wally Schirra, unable to resist temptation, purchased a corned beef on rye from Wolfie's Restaurant in Cocoa Beach, Florida, and presented it to Young while he was suiting up that morning. During the in-flight food test, Young removed the sandwich from his pocket and handed it to Grissom, who took a few bites and joked about it to ground-controllers. Members of the House of Representatives Appropriations Committee were not amused. During subsequent budget hearings, one congressional leader went so far as to suggest that the very thought of an astronaut sneaking snacks onto a spacecraft was "frankly . . . just a little bit disgusting." NASA associate adminis-

Table 2

Project Gemini Manned Missions

Mission	Dates	Crew	Duration (Days:Hours:Minute)
Gemini 3	23 March 1965	Virgil Grissom	0:4:53
		John Young	
Gemini 4	3–7 June 1965	James McDivitt	4:1:6
		Edward White	
Gemini 5	21–29 Aug. 1965	L. Gordon Cooper	7:22:55
		Charles Conrad Jr.	
Gemini 7	4–18 December 1965	Frank Borman	13:18:35
		James Lovell	
Gemini 6-A	15–16 December 1965	Walter Schirra	1:1:51
		Thomas Stafford	
Gemini 8	16–17 March 1966	Neil Armstrong	0:10:41
		David Scott	
Gemini 9-A	3–6 June 1966	Thomas Stafford	3:0:21
		Eugene Cernan	
Gemini 10	18–21 July 1966	John Young	2:22:47
		Michael Collins	
Gemini 11	12–15 September 1966	Charles Conrad Jr.	3:23:17
		Richard Gordon	
Gemini 12	11–15 November 1966	James Lovell	3:22:35
		Edwin Aldrin Jr.	

trator for manned space flight George Mueller nodded in agreement and offered his assurance that steps would be taken "to prevent the recurrence of
corned beef sandwiches in future flights."[5]

For the public the high point of the Gemini program came on 3 June 1965,
when Gemini 4's Ed White opened his overhead hatch, stood up on his seat,
and floated out into space. He was euphoric, but, like Alexei Leonov, he sub-
sequently experienced some difficulty in getting back into the cockpit.
White's crewmate, James McDivitt, had to pull him back down into his ejection seat far enough for them to close the hatch.

Other problems plagued the early Gemini missions. Attempts at rendezvous in space proved difficult. The electricity-producing fuel cells that
would prove so important to the lunar program developed problems. Pin-
point splashdowns continued to elude the Gemini astronauts. Computer fail-

On 3 June 1965 astronaut Edward H. White II became the first American to walk in space. He remained outside Gemini 4 for twenty-one minutes, tethered to the spacecraft by a twenty-five-foot umbilical line. (Courtesy of NASA)

ure was blamed for the fact that Gemini 4 returned to Earth 50 miles (81 kilometers) from a waiting aircraft carrier. Gemini 5 landed 80 miles (129 kilometers) from its target point: a technician had programmed the computer with an inaccurate figure for the diameter of the Earth.

NASA officials were determined to focus on the mechanics of orbital rendezvous and docking. Gemini 6 astronauts Wally Schirra and Tom Stafford spent weeks running through the complex procedures in spacecraft simulators. On 25 October 1965 they were strapped into their seats atop a Titan 2 on pad 19 waiting to chase an Agena target vehicle into space. They watched the Atlas-Agena rocket climb away from pad 14, then waited for confirmation that the Agena upper stage was in orbit. Instead they learned that their mission was scrubbed. Their target had broken up in space.

Schirra and Stafford were still climbing out of their suits when NASA and

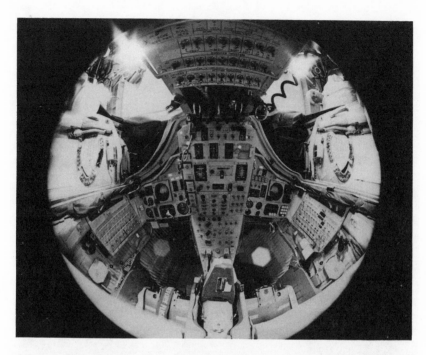

A fish-eye camera lens views the cockpit of Gemini 6. Note the technician peering in the window at the right. Astronauts Wally Schirra and Tom Stafford needed three tries to get the craft into space but finally rendezvoused with Gemini 7 astronauts James Lovell and Frank Borman on 7 December 1965. (Courtesy of NASA)

McDonnell Aircraft officials began to discuss an even more spectacular rendezvous target for the crew of what would become known as Gemini 6-A. Six weeks later, on 4 December, Gemini 7, with James Lovell and Frank Borman aboard, lifted off from the cape. A week after that Schirra and Stafford once again were poised atop a Titan 2, ready to be the next men in space. The booster ignited, then immediately shut down, leaving the two astronauts sitting on a live rocket that had just malfunctioned. By all the rules they should have ejected from the spacecraft, a procedure that would have forced the cancellation of the dual mission. Instead they elected to remain aboard while the engineers shut down all of the systems and retrieved them. The gamble paid off. Three days later, Schirra and Stafford were finally in space and maneuvered to within a few feet of their companions aboard Gemini 7.

Neil Armstrong and Dave Scott of Gemini 8 were the first astronauts to dock with another spacecraft. The procedure did not go well. The two spacecraft began to roll and yaw to the left soon after docking was complete. Armstrong immediately broke the connection and backed away, only to discover that the gyrations were growing worse. We have serious problems here," he informed ground-controllers. "We're tumbling end over end."[6] Armstrong recognized that one of the spacecraft's sixteen maneuvering thrusters was firing continuously. Fighting disorientation and blurring vision as the spacecraft tumbled once a second, he switched the primary maneuvering system off and activated the set of thrusters used to control the craft during reentry. His action worked, and mission planners ordered an immediate return to Earth.

Gemini 6 veteran Tom Stafford held the program record for hard luck. Once again his scheduled launch, this time as commander of Gemini 9, was postponed when his Agena target vehicle was destroyed before it reached orbit. When he and crewmate Gene Cernan finally approached a second docking target in space, they found that the large shroud, designed to protect the docking adapter during launch, was only partially open. Docking, the most important element of the mission, was scrubbed. Cernan was forced to terminate his extravehicular activity (EVA) prematurely when he became so exhausted and overheated that his visor fogged up.

Gemini 10 marked a decided turn for the better. John Young and Mike Collins performed the first completely successful docking in history, then used the Agena motor to transfer into a higher orbit for a rendezvous with the Agena that had been abandoned by Armstrong and Scott. Collins performed two extra vehicular activities (EVAs) and became the first man in history to move between two spacecraft in orbit.

All of the Gemini program goals were achieved during this single mission: rendezvous, docking, maneuvering while docked with a second spacecraft, and an EVA during which useful work was performed. The last two missions went just as smoothly, demonstrating that the problems encountered early in the program had been solved.

The primary goal of Mercury had been to send one American after another into space, capturing and holding public attention while national leaders established an overall goal for the U.S. space program and scientists and engineers determined how that goal might be achieved. Gemini was purposeful. It proved that the technical obstacles to a lunar journey could be overcome, and it provided the means of testing the men and equipment for the journey. The ten Gemini missions represented the first steps on the road to the Moon.

James Lovell *(left)* and Buzz Aldrin return from Gemini 12, the last flight of the Gemini program, on 15 November 1966. Gemini established procedures and provided valuable information for the Apollo missions to come. (Courtesy of NASA)

Just as important, Gemini purchased the time required to prepare the Apollo spacecraft and launch vehicles. By 1966 President Lyndon Johnson had set to work on a series of imaginative and expensive social programs aimed at creating a Great Society. At the same time, the United States was sinking deeper into a Southeast Asian quagmire that would drain the nation of blood and treasure. If the United States, like the Soviet Union, had been forced to stand down from manned space flight for two years, an assassinated president's challenge to go to the Moon might easily have been forgotten. Instead, there was Gemini.

12 • The Apollo Era

Oceans as yet undared my vessel dares;
Minerva lends the breeze, Apollo steers,
And the Muses nine point out to me the Bears.
Dante Alighieri, *The Divine Comedy*

By dawn on 9 November 1967 preparations for the launch of the first Saturn 5 were well underway at Cape Kennedy, formerly Cape Canaveral, Florida. Watching from a point some four miles (six kilometers) from launch pad 39, Mike Collins, now a veteran space voyager, thought that the rocket sitting on the pad looked like "a thick, white pencil, giving off dainty wisps of steam in the thin November sunshine."[1]

In fact, it stood 363 feet (111 meters) tall. Weighing more than 3,000 tons (2,730 metric tons), it was the heaviest object ever to fly. NASA public-relations specialists made it their business to come up with stunning comparisons: one of their favorites was that a Saturn 5 stood taller than the Statue of Liberty and weighed more than a U.S. Navy destroyer. The rocket itself was composed of roughly three million parts. The Apollo command and service module mounted on top contained another two million; the lunar module that eventually would take two astronauts to the surface of the Moon and return them to lunar orbit was built from another one million pieces. As a whole it

The booster for the first flight of the Saturn 5 rolls out of NASA's Marshall Space Flight Center in Huntsville, Alabama, on 27 September 1965. Confident in their technology, and anxious to keep the Apollo program on schedule, NASA officials agreed that the flight itself would serve as the final integration test of the system's more than three million components. (Courtesy of NASA)

was the most complicated machine ever built. With all five of the huge first-stage F-1 engines delivering a grand total of 7.5 million pounds of thrust (34 million newtons), it was also the most powerful flying machine ever crafted by human hands.

In order to produce that much power the first-stage engines consumed 15,000 gallons (56,850 liters) of kerosene fuel and 25,000 gallons (94,750 liters) of liquid oxygen every minute. Fifty-five-thousand-horsepower gas-turbine-driven pumps the size of beer kegs forced 666 gallons (2,524 liters) of propellant into the combustion chambers each second. (Each engine's fuel pump was more powerful than a navy destroyer's four steam turbines put together.) Some of those fuel lines were big enough for a person to crawl through.

As launch time approached, the tension in the control room built to an extraordinary level. Everyone was aware of the fact that this would be the first test for many of the subsystems making up this incredibly complex rocket. A 1963 NASA study had indicated that the program already was falling behind schedule and had suggested that, unless radical steps were taken, the agency had only a one in ten chance of reaching the Moon by the end of the decade. As a result NASA officials had instituted a policy of "all-up" testing for the Saturn 5. Rather than proceeding through rigorous and systematic checks of the individual components and subsystems, the launch itself would serve as an ultimate systems-integration test for the entire vehicle. This approach streamlined the program, but it also left the engineers with a great many unanswered questions. The answers began to pour in at 7:00 A.M. that November day.

As the rocket rose from the pad, tension gave way to awe and wonder. Words cannot adequately describe a Saturn 5 liftoff. You had to be there— and this was a moment that none of those who were would ever forget. "At ignition," Mike Collins recalled, "the flame was orange-red, but rapidly changed to an incandescent white at its core and a dirty brown at its edges. The scene had an eerie quality because for the first 20 seconds it occurred in total silence. When the sound wave reached us with a sudden jolt, it was more than just a noise. The sand under my feet began vibrating and I felt as if a giant had grabbed my shirtfront and started shaking."[2]

The 7.5 million pounds of thrust (34 million newtons) unleashed in those first few seconds was the equivalent of 180 million horsepower. Two million horsepower was experienced as noise. It has been said that the only thing louder than the first-stage ignition of a Saturn 5 is the explosion of a thermonuclear weapon. Some months later, during another Saturn 5 liftoff, television viewers saw the familiar image of Walter Cronkite suddenly blur as the camera began to vibrate. The newscaster put his hands over his head to protect himself from ceiling tiles knocked loose by the rocket's awesome power.

Saturn 501, as the first test was named, was a perfect launch. Each of the three stages fired on cue, sending the CSM into orbit at a speed of 25,000 miles (40,250 kilometers) per hour. Nine hours later the spacecraft splashed down safely in the Pacific recovery zone. Before the Apollo era came to an end a grand total of thirty-two rockets designated Saturn 1, Saturn 1B, or Saturn 5 would fly into space. There would not be a single launch-vehicle failure serious enough to prevent a rocket from performing its mission. Saturn 501 had set the standard for what was to come.

The first Saturn 5 lifts off on 9 November 1967, its engines creating the equivalent of 180 million horsepower of thrust. According to astronaut Mike Collins, watching from four miles away, it "felt as if a giant had grabbed my shirtfront and started shaking." (Courtesy of NASA)

There had never been any illusions about the cost of flying to the Moon. President Kennedy approved the Apollo program on the basis of initial estimates ranging from $20 to $40 billion. For the first few crucial years of the project Congress presented NASA administrator James Webb with what amounted to a blank check. In 1960, the year before President Kennedy presented his space challenge to the nation, the agency had a budget of $523.6 million. Over the next five years that figure increased by more than 1,000 percent, until, in 1965, it reached a record $5.25 billion. Where Apollo was concerned, price was no object.

Not long after his retirement in 1989 Senator William Proxmire (D-Wisconsin), who had earned a well-deserved reputation as watchdog of the nation's budget, was asked why he and his colleagues had funded Apollo with such apparent enthusiasm. "We had a most unusual kind of economic situation," he explained. "One that is in particular contrast with the situation today. . . . We had a very big increase in government revenues because the economy was doing well. And there was a feeling that we wanted to maintain those revenues and not cut taxes. It was argued that what we should do, in order not to slow the economy by running surpluses, was give a substantial amount back through revenue sharing. Therefore, there was funding available to go ahead with this exciting activity in space."[3] Willis Shapley, a budget official responsible for overseeing NASA spending prior to 1965, believed that congressional confidence was well placed. James Webb, he once remarked, "knew what a billion dollars was." In fact, NASA would ultimately bring the Apollo program in slightly under budget.[4]

A rapidly expanding budget was not the only index of NASA's growth. During the boom years, from 1959 to 1967, the agency grew from 9,235 to 35,860 employees, and that was only the tip of the iceberg. At the height of the Apollo program, NASA and its contractors employed some 430,000 individuals. In July 1967 the agency paid $7.3 million for overtime work alone.

Life for those employees was exciting and filled with challenge, but it was seldom easy. "Working for NASA was, at times, hell," admitted George Skurla. In charge of the Grumman Corporation operations at the cape, Skurla recalled that some NASA officials, in their determination to get the job done, could be "demanding and, you might say, arrogant."[5] "I don't think NASA paid for more than about 70 percent of the true human effort that went into Apollo," observed one senior aerospace executive. "A lot of people worked day and night. We all were swept up in it." George Skurla praised the agency for its success in managing an incredibly complex enterprise but noted that "it left a lot of human wreckage in its wake in terms of broken families, divorces, and busted professional careers." And there was not much room for dissent. Those who "gave NASA a hard time about something" were "summarily thrown off."[6]

The pressures that shaped life inside the agency during the Apollo years

were the natural result of a level of activity that can be described only as fre-
netic. NASA was operating a staggering array of overlapping and simultaneous programs, each with its own timetable, problems, demands, dangers,
and constraints.

New families of Earth-orbiting satellites and planetary spacecraft were designed, built, and flown. Officials at the Langley Research Center in Virginia
and the Dryden Flight Research Center in the California desert operated the
world's most extensive flight-test program. In cooperation with North
American Aviation and the U.S. Air Force, it included, for example, a total of
199 flights with the X-15 rocket plane between 1959 and 1968. In Florida six
Mercury astronauts flew into space, and Gemini ran its course, from concep-
tion through the design and construction process, to the completion of ten
successful manned missions.

And then there was Apollo. The lunar missions required an extraordinary
effort involving virtually all of the NASA centers. The agency had to develop
a new set of spacecraft—the command and service module and the lunar
module. Mercury and Gemini had roared aloft atop modified Redstone, Atlas, and Titan 2 missiles. That would not do for Apollo, so NASA developed
the largest and most powerful rockets in history—the Saturn family. Huge
Apollo facilities were constructed near Houston and New Orleans. Cape
Kennedy, as Cape Canaveral was known from 1963 to 1973, doubled in size
as an entirely new Apollo launch complex was built on 126 square miles (324
square kilometers) of wetland adjacent to the old facility.

Goddard Space Flight Center in Maryland, the Jet Propulsion Laboratory
in California, and Langley would be involved as well. Three families of robot
explorers were to blaze a trail to the Moon. The December 1961 announcement that Ranger, a spacecraft with a basic research mission, would be reconfigured to investigate potential lunar-landing sites created a considerable
stir in the space-science community. The growing public perception, rein-
forced by articles in popular magazines and trade journals, that all of NASA's
resources would be dedicated to landing men on the Moon only made matters
worse.

The new office of space science that would function independently of the
manned space operation, and the determination of the director, Homer
Newell, to squeeze the maximum amount of scientific instrumentation into
every unassigned corner of every possible spacecraft, helped reassure space
researchers that they would not be ignored in the Moon race. The fact that sci-

ence would take a back seat to the needs of the Apollo program was, however, perfectly apparent.

At the outset of the U.S. lunar program, the Soviets had already achieved an impressive series of firsts. Luna 2, launched from Baikonur on 12 September 1959, crash-landed on the Moon's Sea of Serenity to become the first manmade object to reach the surface of another world. Luna 3 circled the Moon, transmitting the first pictures of the hidden face of our closest celestial neighbor. In February 1966 Luna 9 made the first successful Moon landing.

In contrast to the Soviet program, which was designed to gobble up any available firsts, NASA officials devised a systematic plan that would answer key questions for Apollo mission planners. Project Ranger, the first step in the program, involved a series of nine spacecraft designed to crash into the lunar surface and send back to Earth photos taken right up to the moment of impact. Ranger began as a hard-luck project, plagued by personnel difficulties and hardware failures. The first six launches (23 August 1961 to 30 January 1964) were failures, but Rangers 7, 8, and 9 (launched on 28 July 1964, 17 February 1965, and 21 March 1965, respectively) performed flawlessly, providing much better images of potential landing spots than could have been obtained in any other way.

The second phase of the unmanned exploration of the Moon involved soft-landing five Surveyor spacecraft on the lunar surface between May 1966 and January 1968. The Soviet spacecraft Luna 9 may have gotten there first, but the Surveyors were part of a more thoroughgoing and sustained program. In addition to proving that the lunar surface would support the weight of a spacecraft, the Surveyors sampled and analyzed the Moon's soil and returned thousands of images of the landscape. In conjunction with the Surveyors, five Lunar Orbiter spacecraft were launched from August 1966 to August 1967. They flew a total of six thousand orbits around the Moon's equator, photographing 99 percent of its near surface. At the conclusion of each Orbiter's mission, it was commanded to crash on the Moon in order to avoid endangering later manned operations.

While the robot explorers were photographing and probing the surface of the Moon for safe landing spots and the Gemini astronauts were testing the equipment and procedures for the lunar missions, veteran rocketeer Wernher von Braun was supervising the construction of the massive Saturns. "Wernher always seemed to be older than he was," Rocketdyne engineer Paul Castenholz recalled. "You looked up to him as having a lot of experience, a very

NASA engineers look for possible Apollo landing sites on a
thirty-foot mosaic of photos taken by Lunar Orbiter 4. The five
Orbiter spacecraft photographed 99 percent of the Moon's visible surface in the course of six thousand orbits. (Courtesy of
NASA)

strong vision."[7] By the mid-1960s that vision had made von Braun the best
known and most visible man in NASA—and one of the most powerful. As director of the Marshall Space Flight Center he stood at the center of all space-
flight activity. Space travel was about rockets, and rockets were Marshall's
business. The astronauts in Houston, the satellites developed at Goddard, the

lunar and interplanetary spacecraft produced at JPL—all of them would remain on the ground without the boosters.

Marshall was the largest NASA center by any measure except square footage. In 1966, the year of peak spending on the Saturn rocket program, Huntsville spent $138.7 million in research and program-management funds alone. That figure was 22 percent of the NASA administrative budget, and was 44 percent higher than the funding level for any other NASA center. And budget figures were not the only measure of Marshall's importance. By 1966, 7,740 individuals, more than 20 percent of the agency's total full-time work force, were employed at Marshall. The center was also responsible for the lion's share of NASA's contractors.

Ultimately, all of the money and all of the people were the responsibility of Wernher von Braun. He was at the peak of his career. "You couldn't help but be impressed by the work he had done," Sam Hoffman, president of Rocketdyne, observed. "He was a leader; he had charisma."[8] NASA officials took full advantage of the power of his image and personality, sending him to testify before Congress time and again. They knew that if anyone had the capacity to light fires of enthusiasm in the hearts of tightfisted and skeptical politicians, it was Wernher von Braun.

For all of the admiration and honors heaped on him, however, von Braun was never able to escape his past completely. One joke circulating during the months following Sputnik suggested there was alarming news from Huntsville. Having given up German for English, so the story went, von Braun was now learning to speak Russian. Perhaps Sam Hoffman summed it up best: "I think he would have gone even further if it hadn't been that he was ... on the wrong side in the War."[9]

Von Braun and his old comrades had shaped Marshall's style and character. In 1960 the center's director, deputy director for research and development, and seventeen of its twenty-three principal office and directorate chiefs were German. Leadership positions held by non-Germans were primarily administrative and nontechnical. Von Braun had urged his team leaders to select American deputies, but the clannish spirit forged at Peenemünde continued to mark the place throughout his tenure as director. As a result there were always ripples of dissent, thinly veiled comments about German "arrogance," and scattered complaints regarding the number of Germans remaining in key positions and the rigid chain of command. James Webb understood the situation and took advantage of it. The Germans, he noted, "were unique in work-

Wernher von Braun and other NASA officials monitor the flight of Saturn SA-8 from the Complex 37 Launch Control Center. Kurt Debus *(pointing)* had been responsible for launching rockets constructed by the von Braun team since the era of Peenemünde. Dr. Eberhard F. M. Rees, standing and leaning into the frame from the right, was another longtime member of the team. Von Braun is seated between Debus and Rees. (Courtesy of the National Air and Space Museum, Smithsonian Institution, Photo No. 91-19622)

ing together, and knew whom they could trust, putting those men who could do best in the job. Those turned out to be the people they had known and worked with and trained in many cases." The success or failure of the Apollo program would rest on the activity at Marshall.[10]

For all of his experience, running the Saturn program presented new problems for von Braun. The scale of the enterprise was staggering, and the old arsenal system was no longer an appropriate management structure. This time he would be functioning as the world's busiest contracting officer. Roger Bilstein, in his official history of the Saturn 5, *Stages to Saturn,* lists 278 major contractors and subcontractors involved in the program and admits that it was "not practical" to present a full list of all of the hundreds of subcontractors.

One experienced administrator referred to the system established at Marshall to manage the Saturn program as "one of the most sophisticated forms of organized human effort that I have ever seen anywhere."[11]

The Saturn program involved the development of three distinct rocket types: the Saturn 1, the Saturn 1B, and the Saturn 5. The two-stage Saturn 1 stood 164 feet (50 meters) tall. The first five rockets in the series, known as Block 1 vehicles, were launched between October 1961 and January 1964. Four of those flights were first-stage tests of the eight Rocketdyne H-1 engines. Not until the fifth launch was a live second stage, powered by Pratt and Whitney's liquid-hydrogen-and-liquid-oxygen-propelled RL-10, ignited in space. The five Block 2 rockets in the series, launched between 28 May 1964 and 30 July 1965, carried test versions of the Apollo command module and three Pegasus micrometeoroid-detection satellites.

The first of the 224-foot (68-meter) Saturn 1B vehicles was launched on 26 February 1966. The Chrysler-built first stage was designated S-1B and was similar to that of the Saturn 1. The second, or S-4B, stage was powered by a single Rocketdyne J-2 engine. In the interest of supporting a vital national program, Pratt and Whitney had shared its expertise in liquid-hydrogen engines with Rocketdyne. The J-2 was far more than an updated version of the tried and true RL-10, however. Capable of pushing a 50-ton (45,500-kilogram) payload to the Moon, the J-2 was the first hydrogen power plant that could be stopped and restarted in space.

In all, nine Saturn 1B rockets would fly into orbit. Declared ready for manned flights after only four tests, the Saturn 1B remained in service until 1975, launching the crews of Apollo 7, the three Skylab missions, and the Apollo-Soyuz Test Project (ASTP).

The Saturn 5 was the ultimate rocket in the series. The first (S-1C) stage, built by Boeing, was powered by five Rocketdyne F-1 engines. North American Aviation, which became North American Rockwell in 1967, was responsible for the second (S-2) stage, which was powered by five J-2s developing 1,150,000 pounds of thrust (5,116,000 newtons). A single J-2 propelled Douglas Aircraft's final S-4B stage out of orbit toward the Moon. Von Braun and his team were directly responsible for designing the instrument unit that sat on top of the S-4B stage.

Thirteen Saturn 5 rockets were launched between 9 November 1967 and 14 May 1973. Ten of those rockets carried people into space. The program was a final technological triumph for the men who had begun their careers in

Rocketdyne J-2 engines in various stages of completion at the company's California assembly plant in July 1966. The J-2 was the first hydrogen engine that could be stopped and restarted in space. (Courtesy of NASA)

Germany so many years before. By 1973 most of the veterans of Peenemünde had either left NASA or were approaching retirement.

The Grumman Corporation won the basic contract for the design and production of the LM that would carry the astronauts down to the Moon from lunar orbit and lift them back again. To almost everyone's surprise, North American submitted the winning bid for the CSM. Both companies were beset by myriad problems, most of which could be traced to the fact that design had begun before the mission plan was fully developed. North American Rockwell ultimately was forced to produce two versions of the CSM, one for testing in Earth orbit, the other a vehicle fully configured for a trip to the Moon. Grumman engineers were pushed to strip 2,400 pounds (1,080 kilograms) of weight from their original LM design. Both Grumman and North American watched helplessly as NASA accountants replaced their profitable cost-plus contracts with incentive agreements designed to force them to meet rigid time and quality requirements.

Disaster struck when it was least expected, not with a crew halfway to the Moon but during a routine ground test at the start of the program. Just after lunch on the afternoon of 27 January 1967 astronauts Virgil Grissom, Edward White, and Roger Chaffee, the first crew scheduled to fly Apollo into space, took the elevator to the top of the launch pad 34 service tower at Cape Kennedy, walked through the white room, and climbed into the Apollo-Saturn 204 command module. Five hours later, they were still sealed in the spacecraft, practicing flight procedures while struggling to correct a communications problem.

At precisely 6:31 P.M. an alarmed Gus Grissom yelled over the radio, "There's a fire in here!" Smoke, heat, and fumes erupted into the white room, preventing technicians from approaching the capsule. A team of workers had the fire under control and the hatch open within five minutes, but it was too late. The three astronauts were dead from asphyxiation and burns.

The Apollo-Saturn 204 catastrophe was not a freak accident. In the fire's aftermath it is difficult to imagine how the danger could have been overlooked. The life-support system supplied a pure-oxygen environment, at atmospheric pressure, to a capsule filled with flammable materials and hundreds of potential ignition sources. The new spacecraft had been the proverbial accident waiting to happen.

In fact, the dangers had not gone unnoticed. Gen. Sam Phillips, an air force officer assigned to assist NASA with Apollo planning, had issued a series of stinging reports complaining of poor workmanship on the part of several contractors. An employee of North American Aviation had been fired for calling attention to safety problems related to the command module. Just three weeks before the fire, astronaut physician Charles Berry complained of the specific hazards faced during ground tests, when the cabin pressure was much higher than it would be in space and the danger of fire much greater.

Suddenly the very future of Apollo was in danger. Without the presence of

James Webb, who worked to restore confidence and to get the program back on track, the Moon effort might have come to an end. Webb's first step was to establish a NASA accident-investigation board headed by Floyd Thompson, director of the Langley center. He then faced Congress, the White House, and the public, explaining and defending his program and justifying the allocation of an additional $50 million to redesign and rebuild the Apollo spacecraft.

The increasing pressure to move forward, coupled with the incredible complexity of the hardware required to transport people to the Moon, had forced safety and quality assurance into the background. That could never happen again. New safety procedures were instituted, and NASA employees were forced to remain absolutely current on the status of every aspect of spacecraft and booster development. At the same time, the Apollo spacecraft was checked, rechecked, and redesigned. NASA investigators compiled an initial list of 8,000 potential problems that had to be resolved. Ultimately, 1,697 changes were recommended to the NASA configuration control board that was established to make the final design decisions. The board approved a total of 1,341 alterations.

The bitter lessons of the Apollo-Saturn 204 disaster would eventually be ignored, but not before men reached the Moon. The period of thoughtful reassessment and redesign following the fire gave hard-pressed engineers some welcome breathing room. The pressure for immediate flights was relaxed. There was time to catch up with a program that had been racing ahead so rapidly that no one could keep track of all the details.

Following the death of Sergei Korolev, Vladimir Chelomei and Vasili Mishin, who inherited Korolev's bureau, emerged as the leading figures in the Soviet space effort. Chelomei, who spent most of his career developing cruise missiles and ICBMs, now was responsible for both of the large boosters that would be critical to the success of any lunar program. The 196-foot (59.8-meter), three-stage Proton was designed to support manned circumlunar flights, whereas the 370-foot (113-meter), four-stage SL-12 rocket (type N or type G), like the Saturn 5, would boost a lunar-landing mission into space.

By the fall of 1966 it became apparent that the first Proton SL-12 test flights were still months away. In an effort to return to manned space flight it was decided to proceed with the first unmanned orbital tests of the new Soyuz craft, using an updated version of Korolev's Voskhod SL-4 (R-7) booster. The initial launch of the new Soyuz spacecraft/rocket combination came on 28 November 1966. Called Cosmos 133, the spacecraft was recovered after two days in orbit. The heat shield had failed, allowing superheated plasma to sear and score the pressure hull. On 7 February 1967 a further Soyuz flight test, Cosmos 140, revealed additional problems with the attitude thrusters and temperature-control system. Incredibly, Moscow insisted that the first manned launch proceed as scheduled. Vasili Mishin initially refused to cooperate, but he was overruled by his superiors.

Cosmonaut Vladimir Komarov began his second space mission early on the morning of 23 April 1967, four months after the Apollo fire. At forty he was the oldest cosmonaut, and the first to fly into orbit since the launch of Voskhod 2 more than two years earlier. Evidence indicates that the Soviets were planning a spectacular return to space. Komarov would be flying Soyuz 1, the first of a new generation of Soviet spacecraft. Designed to carry a crew of cosmonauts around the Moon, Soyuz consisted of three modules: at the base of the craft was a cylindrical propulsion-and-instrument module, the crew rode into orbit and returned to Earth aboard the central descent module, and the nose of the spacecraft was a spherical orbital module with work and experimental areas.

Cosmonauts Valery Bykovsky, Yevgeny Khrunov, and Alexei Yeliseyev most likely were scheduled to follow Komarov into orbit aboard Soyuz 2. Presumably, the two spacecraft would conduct the first rendezvous and docking in the history of the Soviet space program. But problems with Soyuz 1 began soon after launch. A solar panel failed to deploy, depriving Komarov of electrical power. Then the thruster-control system malfunctioned. Struggling to control the ship's gyrations, Komarov was ordered to nurse his malfunctioning spacecraft back to Earth. Reentry attempts on the sixteenth and seventeenth orbits failed, apparently because Komarov could not stabilize Soyuz 1 for retrofire. Orbit eighteen offered the last opportunity to bring the craft down in the Soviet Union that day. Komarov made the attempt, but when the tumbling spacecraft fouled its recovery parachute, Soyuz 1 hit the ground at 310 miles (500 kilometers) per hour. Vladimir Komarov earned the dubious honor of becoming the first person to die during the course of an actual spaceflight.

There were no more Soviet manned missions for the next eighteen months. When flights resumed they were Earth-orbital missions with goals apparently similar to those of the Gemini program. On 26 October 1968 Georgi

Beregovoi brought his Soyuz 3 spacecraft within 650 feet (200 meters) of the unmanned Soyuz 2 vehicle launched a day earlier.

Only ten weeks later, on 16 January 1969, Vladimir Shatalov, piloting Soyuz 4, docked with Soyuz 5, which was crewed by Boris Volynov, Yevgeny Khrunov, and Alexei Yeliseyev. Khrunov and Yeliseyev performed the first spacewalk crew transfer, returning to Earth on Soyuz 4. The Soviets operated three crews (Soyuz 6, 7, and 8) in orbit from 13 to 16 October 1969. Soyuz 9 (Andrian Nikolayev and Vitaly Sevastyanov), the last crew to fly before the Salyut space station went into operation, remained aloft for nearly eighteen days, from 1 to 19 June 1970, breaking the previous record for time spent in Earth orbit set by Gemini 7.

The early Soyuz missions provided an opportunity for spacecraft rendezvous and docking in Earth orbit. During the same period, seven unmanned Zond spacecraft were launched by Proton SL-12 boosters. Cosmos 146 and 154 and Zond 4 to Zond 8 were stripped-down, modified versions of the Soyuz instrument and descent modules. Zond 1, 2, and 3 had been unrelated and much smaller interplanetary probes launched between 1964 and 1965. Zond 4, possibly the first Moon-shot test, entered a highly eccentric Earth orbit. Zond 5, launched on 14 September 1968, flew around the Moon and returned to Earth, splashing down in the Indian Ocean. Clearly, Zond 5 was the first step in a program to send cosmonauts around the Moon, but it had encountered reentry G-loads that were much too high for a human crew to survive. Zond 6, launched on 10 November 1968, returned to Earth using a different reentry technique that seems to have solved the problem.

The flights of Soyuz 3 and Zond 6 aroused worry in NASA officials about the possibility of the Soviets capturing the honor of a manned circumlunar mission from Apollo 8, which was scheduled for December 1968. In fact, the next Zond flight did not occur until August 1969, a month after the first U.S. Moon landing. Phillip Clark, a British student of the Soviet space program, has suggested that vibration problems with the Proton launcher prevented the manned attempt.

The Soviets seem to have continued work on their lunar program at least through 1972, pinning their hopes on the SL-12 (type N or type G) booster. One of the first of these giants, along with its pad, was destroyed in an explosion on 3 or 4 July 1969. Another rocket actually was launched during 23–24 June 1971, its thirty main engines developing 10 million pounds of thrust (44,483,985 newtons), but it disintegrated before reaching orbit. The final booster in the series suffered a multiple engine failure short of orbit on 24 November 1972.

The Soviets did not go to the Moon with a manned vehicle. Twenty years later, however, much of the hardware developed to accomplish that task remains in service. The Soyuz spacecraft shuttles cosmonauts back and forth from Earth-orbiting space stations, permitting the Soviets to come close to achieving a goal that has eluded their American rivals: maintaining a long-term presence in space.

The Proton also has become something of a workhorse—and, ironically, a pioneer of space capitalism. For a price, the Soviets will send any reasonable payload into orbit aboard either a Proton SL-12 or SL-13 booster. All things considered, the story of the effort to send cosmonauts to the Moon does not have such an unhappy ending after all.

13 • Men from Earth

Here men from the planet Earth
First set foot upon the Moon
July 1969, A.D.
We came in peace for all mankind

Robert Seamans—NASA associate administrator, 1960–65; NASA deputy administrator, 1965–68; and secretary of the air force, 1969–73—was a man accustomed to meeting with presidents. Nevertheless, he has particularly vivid memories of an encounter with Lyndon Johnson prior to the first Apollo flight in 1967. The president called the meeting to impress cabinet officers and agency heads with the need for an across-the-board 5 percent budget reduction. "He gave us a big lecture in straightforward, barnyard language about cutbacks," Seamans recalls. At the conclusion, Johnson announced that he would stand at the door to shake hands with each of those attending and receive a personal commitment to make the required cut. Seamans agreed that NASA would do its best to comply, adding that he was sure the president "wouldn't want us to jeopardize the safety of the astronauts." As he was leaving the room he overheard the president instruct an aide to "tell that young guy the facts of life."[1]

As a U.S. senator and as vice president Johnson had been one of the most important political supporters of a strong national space effort. As president his priorities shifted. Determined to both fund a series of major social programs aimed at building a Great Society and fight an enormously expensive war in Southeast Asia, Johnson could no longer justify supporting NASA at the level of the peak Apollo years.

The NASA budget had increased every year from 1959 to 1965. The beginning of a steady decline that would continue for ten years had come in 1966, when for the first time in the history of the agency the budget was lower than that of the previous year. The number of NASA employees rose to 35,860 in 1967 and has declined every year since. The first lunar landing was still two years away, but the flush times were over for the nation's civilian space agency.

James Webb put NASA on the road to the Moon, but he did not stay to enjoy the moment of triumph. A Kennedy appointee, he enjoyed a cordial but not particularly close relationship with President Johnson, who had installed one of his own, Thomas O. Paine, as deputy administrator. During a visit to the White House in September 1968 Webb suggested that he might ask to retire in the not too distant future. He was stunned when the president accepted his "resignation" on the spot.

Three decades after he left office, James Edwin Webb remains the most important single figure in NASA's history. When he came to the agency in 1961, it was relatively small and struggling to keep up with the Soviets in space. During the seven years of his administration, NASA became one of the fastest growing and most visible of the federal government's agencies. Webb the technocrat succeeded in uniting the forces of government and industry to achieve a great national goal. At the height of its activity, the civilian space agency supported 430,000 people, most of whom worked for the businesses, universities, and industrial firms holding NASA contracts or subcontracts. The agency funded research and development in the most advanced areas of technology, including computers and electronics, energy, metallurgy, materials processing, and chemistry.

By the spring of 1968 NASA was ready to send Americans back into space. A total of seventy-three men were selected as astronauts between 1959 and 1975. Forty-three of them would fly into space before the Apollo era came to an end with the July 1975 Apollo-Soyuz Test Project. Twenty-nine of those astronauts would fill the thirty-three crew slots available on the Apollo 7 to Apollo 17 missions. Twenty-four of them would orbit the Moon, and twelve would walk on its surface.

For the most part, the Apollo astronauts were first-born sons (twenty-two out of twenty-nine) from the American heartland. Four states—Ohio, Texas, New Jersey, and Illinois—produced three astronauts each. The South and the West Coast tied with two apiece. Only one astronaut was a New Englander. More of them were born outside the United States (two) than in New York City (zero). As might be expected, almost half of the Apollo crewmen were educated at the U.S. military academies. Annapolis beat out West Point, eight to five. Purdue and the University of Colorado produced two apiece. Pete Conrad, a Main Line Philadelphian and a Princeton alum, was the only Ivy Leaguer in the bunch. Edwin "Buzz" Aldrin Jr. (MIT) came into the program with a Ph.D. All of them, with the exception of geologist Harrison Schmitt, were professional aviators. Some of the Apollo astronauts came to NASA directly from one or another of the abortive military manned space programs—Blue Gemini, the Manned Orbiting Laboratory (MOL) Program, or, as in Neil Armstrong's case, Dyna-Soar.

The average Apollo crewman stood just under 5 feet 10 inches (178 centimeters), weighed 160 pounds (72 kilograms), and was 38.6 years old at the time of his first flight. At 40, veteran Al Shepard was the oldest man to set foot on the Moon. They tended to have blue eyes (sixteen), brown hair (nineteen), and high intelligence. The IQ scores of the original seven astronauts ranged from 130 to 145. The eleven members of the sixth and final Apollo-era selection group had a mean IQ of 141.

They were family men: Harrison Schmitt, Thomas Mattingly, and Jack Swigert were the only bachelors. The average astronaut family included 2.6 children. Twenty-three of them were Protestants; six were Catholics. Several of them were inspired to make spiritual gestures while their missions were underway. Frank Borman read a passage from Genesis while circling the Moon on Christmas Eve 1968. (An irate atheist sued NASA and lost.) Buzz Aldrin administered communion to himself in the cramped quarters of a lunar module resting on the Sea of Tranquility. Edgar Mitchell conducted an experiment in telepathy with friends on Earth and charged that those who doubted the positive results of his tests "did not understand statistics."[2] Mike Collins was universally accorded the title of best handball player. Any contest to select the best comedian in the group would have been hard fought, but Pete Conrad surely would have finished near the top. During the process of astronaut selection a psychologist presented him with a plain white sheet of paper and asked him what he saw. Conrad smiled and told the fellow he was holding it upside down.

Initially mission planners considered sending the first human beings to the Moon in four steps: Apollo 7 (Grissom, White, and Chaffee were the crew of Apollo 1, and five unmanned launches followed) would test the command and service module in Earth orbit. Apollo 8 would operate the first lunar module in orbit and test rendezvous and docking procedures. Apollo 9 would fly into lunar orbit, at which point the mission commander and LM pilot would enter the LM, drop toward the surface of the Moon, and return to dock with the CSM. If all went well, the crew of Apollo 10 would have the honor of being the first human beings to walk on another world.

The first of the redesigned Block 2 Apollo command modules arrived at Cape Kennedy in May 1968 with a message from North American Rockwell stenciled on the crate: "We care enough to send the very best." The crew of Apollo 7 flew the CSM into Earth orbit on 11 October 1968. Astronauts Walter M. Schirra (the only man to fly in the Mercury, Gemini, and Apollo programs), Donn F. Eisele, and R. Walter Cunningham remained in space for ten days and twenty hours, completing twenty-three Earth orbits, beaming the first television pictures to Earth from a manned spacecraft, and turning in a generally flawless performance. The CSM operated as advertised.

Grumman, plagued by design and production problems stemming from the need to reduce weight to an absolute minimum, was running behind schedule on the delivery of the first LM. The recent flurry of Soyuz missions operating in Earth orbit, however, as well as the flight around the Moon of three unmanned Zond spacecraft, clearly suggested that the Soviets were planning to upstage their U.S. rivals by sending a crew of cosmonauts on a circumlunar voyage. Rather than standing down while they waited for the LM, NASA officials decided to insert just such a mission into their own program.

James McDivitt, originally scheduled to command Apollo 8, became such an authority on the LM that he requested the reassignment of his crew to the Apollo 9 mission, which now would have the task of testing the first LM in space. Frank Borman, the designated Apollo 9 commander, agreed to fly Apollo 8 with his crew, Mike Collins and William Anders. When Collins was unable to fly for medical reasons, Jim Lovell stepped in to replace him.

With its new mission, Apollo 8 held the attention of the public until the LM was ready to fly and provided some of the most unforgettable moments and images of the entire lunar program. Borman, Anders, and Lovell set out on the first voyage out of Earth orbit on 21 December 1968 and spent twenty hours circling the Moon. The high point of the mission, and one of the high points of the entire Apollo program, came on Christmas Eve, when millions of people around the globe watched a special television broadcast featuring the three astronauts in lunar orbit. The crew of Apollo 8 described their mission and offered viewers a glimpse of the lunar surface sweeping past the window of the spacecraft. Then commander Frank Borman read the first ten verses of the Book of Genesis and sent holiday greetings back to everyone "on the good Earth."

Not since the flight of John Glenn, almost seven years earlier, had Americans felt such enthusiasm for the space program. Human beings had yet to step on the Moon, but they had journeyed there and returned home safely, with pictures to prove it. The enduring legacy of Apollo 8 is an image of our home planet, a lovely blue-green sphere suspended above the harsh lunar landscape. Set against the blackness of space, the earth looks small, delicate, and incredibly inviting. If that photo did not inspire the environmental movement, it certainly fueled growing concerns over the irreversible effects of human civilization on the earth. Seldom has a single photograph had so profound an impact, reminding everyone of the fragility of our planet and the need for cooperation among all those who call it home.

Two more missions followed in quick succession. James McDivitt, David Scott, and Russell Schweickart, the crew of Apollo 9, checked out all the hardware, including the lunar module, in Earth orbit. Apollo 10 went back to the Moon, where commander Thomas Stafford and pilot Eugene Cernan flew the LM to within 50,000 feet (15,240 meters) of the surface and then returned to a safe docking with John Young in the command module. Just eight years earlier President John Kennedy had challenged the nation to journey to the Moon before the end of the decade. Now the stage was set for the great event.

The men of Apollo 11—commander Neil Armstrong, LM pilot Buzz Aldrin, and CSM pilot Mike Collins—came by their historic roles through the natural process of crew rotation. Acutely aware that all fighter pilots are, by nature, highly competitive, NASA went to extraordinary lengths to keep antagonism over crew assignments to a minimum. At the start of the Apollo program six mission commanders were selected on the basis of their flight experience and seniority: Schirra, Borman, McDivitt, Stafford, Armstrong, and Conrad. The remaining astronauts were assigned as either LM or CSM pilots under one of the commanders. Both a prime and a backup crew were specified in each of the missions being planned. The backup crew stayed abreast of the specific requirements and procedures to be followed by the prime crew and

Views of Earth, such as this one taken in December 1972 from Apollo 17, the last lunar mission, underscored the unique beauty and fragility of our environment set against the void of space. (Courtesy of NASA)

were prepared to step in should an injury or illness occur at any point during training. The backup crew trained to serve as the prime crew three flights after their backup duty.

Armstrong, assigned to back up Frank Borman and the original Apollo 9 crew, should have commanded Apollo 12. When the missions were shuffled, however, the backup followed the prime. Borman accepted the shift to Apollo 8, and Armstrong in turn became the commander of Apollo 11. Aldrin was Armstrong's original LM pilot, with Jim Lovell serving as CSM pilot. When Borman's CSM pilot, Mike Collins, withdrew from Apollo 8 for spinal surgery, Jim Lovell replaced him. Collins returned as Lovell's substitute on the crew of Apollo 11. By the luck of the draw Armstrong, the most private of all the astronauts, would be watched by millions as he walked on the Moon.

The most experienced pilot among the astronauts, Neil Armstrong had learned to fly before he earned his driver's license. Growing up in Wapakoneta, Ohio, he had had a recurring dream. "I could, by holding my breath, hover over the ground," he recalled many years later. "I neither flew nor fell. I just hovered. There was never any end to the dream." Jacob Zent, a local amateur astronomer, cherished the memory of nights when the young neighbor boy joined him at the telescope to "look and look and look."[3]

Armstrong's father took two-year-old Neil to the National Air Races in Cleveland. By the time the boy entered his teens he was an experienced modeler, with a homemade wind tunnel in his basement. A classic "airport kid," he started flying at fifteen, paying for lessons by doing odd jobs around the field and anywhere else he could find them.

During the Korean War, Armstrong served as a naval aviator, flying seventy-eight combat missions from the deck of the USS *Essex* and on one occasion ejecting after losing a wingtip to an antiaircraft cable. The young veteran returned home, earned a degree in aeronautical engineering from Purdue University, and in 1955 became an engineering test pilot flying the North American F-100, the Lockheed F-104, and the Bell X-1B for the National Advisory Committee for Aeronautics (NACA). By 1960 he was widely recognized as one of the best in the business and was assigned to the North American X-15 program. Over the next two years Armstrong would push that aircraft to speeds up to 3,989 miles (6,422 kilometers) per hour and altitudes as high as 207,500 feet (63,246 meters).

In the opinion of the late NASA test pilot Milt Thompson, "Neil was probably the most intelligent of all the X-15 pilots, in a technical sense." Like most of the Edwards Air Force Base fraternity, he also knew how to have a good time with friends, one of whom remembers him as "a pretty good piano player" who "could be coaxed into playing some songs" when the mood suited him.[4] At a more fundamental level Armstrong was a careful, quiet man who liked to think things over before offering an opinion. Mike Collins thought that Armstrong "savored" decisions, "rolling them around on the tongue like a fine wine and swallowing at the very last minute."[5] Many colleagues found him a bit distant. "I knew Neil," remarked Milt Thompson, who worked and flew with him for more than six years, "but I did not know him."[6]

A very different man would be standing next to Armstrong in the cramped cabin of the LM. Like Collins, Buzz Aldrin came from a military family. His father was a pioneer military aviator who had served as an aide to Gen. William "Billy" Mitchell, had known Orville Wright, and had flown with Charles Lindbergh *and* Jimmy Doolittle. Buzz, whose nickname had been supplied by an older sister unable to pronounce her brother's—or "buzzer's"—name, was an all-around athlete with a special gift for science and math. He graduated from West Point, earned his wings as a USAF fighter pilot in 1952, and scored two victories flying a North American F-86 against North Korean MiGs. After the war he gave up an opportunity to attend test-pilot school in order to earn a doctorate at MIT.

Aldrin concentrated his graduate studies on celestial mechanics. By the time he was accepted as a member of the third astronaut group he had become one of the nation's authorities on the intricacies of orbital rendezvous. He not only spoke the language of the NASA engineers, he could correct their textbooks. And he never allowed his colleagues to forget that he was the expert. His attempts at informal education did not always go well and earned him a second nickname: "Dr. Rendezvous." At one point during the planning for Apollo 8 he approached commander Frank Borman with the latest information on orbital maneuvers. Borman's response was immediate: "Goddamn it, Aldrin, you got a reputation for screwing up guys' missions, and you're *not* going to screw up mine!"[7]

Aldrin was a competitor's competitor. Learning that the Gemini astronauts would be given special consideration for upcoming Apollo flight assignments, Aldrin forcefully reminded Deke Slayton of his special credentials and petitioned for a Gemini slot. He soon discovered that "the stick-and-rudder guys in the astronaut office" regarded his direct approach as bad form. Slayton assigned him to Gemini 13, a mission that everyone knew would not fly. But when Charles Bassett and Elliott See, who were scheduled to fly Gemini 9, died in a plane crash, Aldrin dropped into the Gemini 12 crew. It would not be the last time that the sacred and unalterable rules of crew rotation worked in his favor.

Discovering in the spring of 1969 that the crew of Apollo 11 would be the first to land on the Moon, it occurred to Buzz that he might be able to set foot on the lunar surface before the commander. In Gemini the men in the right-hand seats—the copilot position—had made the space walks; perhaps that should also be the case in Apollo. Perhaps the commander should remain on board the LM, at least for a time, to handle potential emergencies. He broached the subject with Armstrong, who replied, in a restrained but firm tone, that he was "not going to rule himself out of some historical position." Aldrin then took the matter to NASA deputy administrator George Low, who ruled in Armstrong's favor. Ten years later a bemused Mike Collins noted in *Carrying the Fire* that "fame has not worn well on Buzz. I think he resents not being the first man on the moon more than he appreciates being second."[8]

The crew of Apollo 11 blasted away from Cape Kennedy's pad 39A on 16 July 1969. Three days later they entered lunar orbit. "Despite the fact that I have spent years studying photographs from Ranger, Lunar Orbiter and Surveyor, as well as from Apollo 8 and 10, it is nevertheless a shock to actually see the moon firsthand," Collins wrote.

The first thing that springs to mind is the vivid contrast between the Earth and the moon. . . . I'm sure that to a geologist the moon is a fascinating place, but this monotonous rock pile, this withered, sun-seared peach pit out my window offers absolutely no competition to the gem it orbits. Ah, the Earth, with its verdant valleys, its misty waterfalls. . . . I'd just like to get our job done and get out of here.[9]

The drama built toward a climax as Armstrong and Aldrin climbed into the lunar module Eagle and separated from the CSM. A 29.8-second burn of the descent engine dropped the LM from the CSM's 60-mile (97-kilometer) altitude down into an orbit that would carry them to within 8 miles (13 kilometers) of the surface. With everything in order, Houston authorized powered-descent initiation (PDI). From that moment the crew of the LM had twelve minutes to touch down on the Moon.

Armstrong and Aldrin made the trip standing side-by-side in the narrow cockpit, fully clad in pressure suits and helmets and tethered to the floor by elastic restraint cords. At the outset of PDI they actually were lying on their backs, flying parallel to the surface. Gradually, the LM began to pitch over until it was dropping straight down. Buzz kept his eyes on the instruments, calling out the readings to Armstrong, who studied the surface through a small, triangular window at his side. As they passed through 4,000 feet (1,828 meters), flight director Gene Cranes gave final approval for a landing. In fact, this trip did not have a "point of no return." During the descent, they were burning the rocket engine on the base of the LM. In the event of a problem they could always drop the descent stage and climb back up to the CSM, using the ascent motor designed to lift them off the Moon at the end of the mission. Of course, if the ascent motor failed to ignite at that point . . .

The computer did most of the flying on the way down, signaling a data overload several times during the final minutes. Radar showed that the de-

AS-506 carries the crew of Apollo 11—Neil Armstrong, Buzz Aldrin, and Mike Collins—from launch pad 39A on 16 July 1969. (Courtesy of NASA)

scent was going according to plan, however, and Armstrong continued to punch the instruction "proceed" into the keyboard. By the time they reached an altitude of 700 feet (213 meters) above the Moon's surface it was obvious that the computer was about to land them in a field of boulders at the edge of a crater. Armstrong engaged his hand controller, thus taking over piloting duties from the machine, and while Aldrin called out altitude and horizontal speed readings, the man who once had dreamed of "just hovering" guided the spacecraft forward and allowed it to drop very slowly as he searched for a landing spot. They finally came to rest on the Sea of Tranquility, with roughly twenty seconds' worth of propellant remaining in the tanks. "Houston," Armstrong radioed, "the Eagle has landed."

Six hours later Neil Armstrong stepped onto the Moon. After Aldrin joined him the two astronauts planted the U.S. flag, set up a group of scientific instruments that would continue to send information back to Earth long after they had departed, and spoke via a special telephone link with President Richard Nixon. The rest of the mission—liftoff from the Moon, rendezvous with the CSM in lunar orbit, the return to Earth, reentry, and splashdown— went precisely according to plan. The first manned flight to the Moon was a textbook mission.

Six successful NASA expeditions to the Moon followed Apollo 11. Twelve astronauts would spend a total of three hundred hours on the surface. They probed, poked, prodded, and drilled the lunar landscape, and they took some of the most spectacular snapshots in the history of photography.

Every spare moment on the Moon was devoted to science. Armstrong and Aldrin were given a relatively simple experimental program. They set up an aluminum-foil apparatus that enabled scientists to determine the elemental and isotopic composition of the solar wind, installed four seismometers for recording moonquakes and micrometeoroid impacts, and positioned an array of fused silica cubes to reflect back to Earth a laser beam that provided a precise measurement of the distance between the two bodies as well as additional clues about the Moon's structure.

Beginning with Apollo 12 each mission left an automated, scientific minilaboratory called an Apollo Lunar Surface Experiments Package (ALSEP) on the Moon. Each ALSEP contained a different set of instruments designed to both answer specific questions and return information of general interest to geophysicists. Powered by a radioisotope thermoelectric generator (RTG) and furnished with a transmitter capable of returning to Earth up to nine million instrument readings a day, each of the five ALSEPs had a predicted life of one year. The Apollo 12 unit, however, was still broadcasting five years after it began operation.

The Apollo crews devoted most of their active hours on the surface to lunar field geology. The missions brought home a total of 842 pounds (379 kilo-

Astronaut Buzz Aldrin poses for a lunar snapshot in July 1969. Photographer Neil Armstrong can be seen reflected in his visor. (Courtesy of NASA)

grams) of Moon rock. The crew of Apollo 11 remained on the surface for the shortest period of time, transporting back to Earth only 46 pounds (21 kilograms) of sample material. Gene Cernan and Harrison Schmitt of Apollo 17 spent three days on the Moon and returned with 243 pounds (109 kilograms). The material ranged from surface dust to basaltic lava samples created during

HORNET + 3

President Richard Nixon congratulates the crew of Apollo 11 during their temporary quarantine aboard the recovery ship USS *Hornet* following their return from the Moon. (Courtesy of NASA)

a partial melting of the lunar surface 3 to 4 billion years ago to beads of ancient volcanic glass and rocks from the lunar highlands that dated to the earliest history of the Moon, some 4.6 billion years ago. Pete Conrad and Alan Bean, the surface crew of Apollo 12, brought back some additional treasure from the Ocean of Storms: bits of metal, electronic components, and optical equipment from Surveyor 3. Scientists and engineers were interested in studying the character of materials that had "soaked" in the lunar environment for two years.

Although Apollo 12 was meticulously planned, using detailed photos of the potential landing site taken by a Lunar Orbiter spacecraft, the difficulty of landing reasonably close to a small robot probe was obvious. The astronauts thought it was a joke when, in order to avoid blowing Surveyor over or covering it with a fresh layer of lunar dust, geologists cautioned them not to land within 500 feet (152 meters) of it. At no time during their descent did the crew see the Surveyor, nor was it visible from the windows of the LM as they were sitting on the surface. When Bean climbed down the ladder and looked

Table 3

Apollo Manned Missions

Mission	Dates	Crew	Duration (Days:Hours:Minutes)	Mission Objective
Apollo 7	11–12 October 1968	Walter Schirra Donn Eisele R. Walter Cunningham	10:20:9	Earth orbit
Apollo 8	21–27 December 1968	Frank Borman James Lovell William Anders	6:3:1	Lunar orbit
Apollo 9	3–13 March 1969	James McDivitt David Scott Russell Schweickart	10:1:1	Earth orbit
Apollo 10	18–26 May 1969	Eugene Cernan John Young Thomas Stafford	8:0:30	Lunar orbit
Apollo 11	16–24 July 1969	Neil Armstrong Edwin Aldrin Jr. Michael Collins	8:3:9	Lunar landing
Apollo 12	14–24 November 1969	Charles Conrad Jr. Richard Gordon Alan Bean	10:4:36	Lunar landing
Apollo 13	11–17 April 1970	James Lovell Jack Swigert Fred Haise	5:22:55	Lunar landing (aborted)
Apollo 14	31 Jan.–9 Feb. 1971	Alan Shepard Stuart Roosa Edgar Mitchell	9:0:20	Lunar landing
Apollo 15	26 July–11 August 1971	David Scott James Irwin Alfred Worden	12:7:12	Lunar landing
Apollo 16	16–27 April 1972	John Young Thomas Mattingly Charles Duke	11:1:51	Lunar landing
Apollo 17	7–19 December 1972	Eugene Cernan Ronald Evans Harrison Schmitt	12:13:52	Lunar landing

around, he was astounded to see the little spacecraft sitting just 600 feet (183 meters) away.

The astronauts of Apollo 11 and Apollo 12 did their exploring on foot, carrying all their equipment in their hands. Tipping the scales at more than 300 pounds (135 kilograms) on Earth, fully equipped Apollo astronauts weighed less than 60 pounds (25 kilograms) on the Moon. But the business of moving about the surface clad in a cumbersome pressure suit and helmet that offered less than perfect mobility and visibility proved more difficult than imagined. Even the simple act of dumping a scoop of lunar soil into a sample bag could be full of surprises, as Pete Conrad explained: "First you had to handle the shovel differently, stopping it before you would have on Earth and tilting it to dump the load much more steeply, after which the whole sample would slide off suddenly."[10]

Apollo 14 commander Alan Shepard, the only one of the original seven Mercury astronauts to walk on the Moon, and colleague Ed Mitchell were provided with a modularized equipment transporter (MET), a sort of two-wheeled lunar rickshaw designed to carry tools, cameras, scientific equipment, and samples. The last three Apollo missions were equipped with lunar rovers, stripped-down electric automobiles featuring webbed "lawn-chair seats" for the two astronauts and an assortment of high- and low-gain antennas and television cameras that enabled them to remain in touch with Houston. With an Earth weight of some 462 pounds (208 kilograms), the rovers tipped the scales at only 77 pounds (35 kilograms) on the Moon. Each of their four wire-mesh wheels was powered by a separate electric motor. Stowed folded during the flight in a compartment of the LM descent stage, the little vehicles greatly extended the mobility and range of the surface parties.

Harrison Schmitt noted that climbing aboard a rover was considerably more difficult than entering a car. "You stand facing forward by the side of the vehicle, jump upward about two feet with simultaneous sideways push, kick your feet out ahead, and wait until you slowly settle into the seat, ideally in the correct one." Somehow, each one of the crews managed to break one or more of the fragile plastic fenders. Discarded maps were clamped in place to prevent the astronauts from being completely covered by lunar dust thrown up by the wheels.[11]

All overland treks with the rovers were limited to a radius of six miles (ten kilometers) from the LM, the maximum distance a pair of tired and worried astronauts could reasonably be expected to walk back from a disabled vehicle. The rover could negotiate slopes of up to twenty-five degrees and had an average speed of seven miles (eleven kilometers) per hour. The Apollo 17 astronauts set an unofficial lunar speed record of eleven miles (eighteen kilometers) per hour while rushing down a slope. By the end of their mission they had set another unofficial record by putting twenty-two miles (thirty-five kilometers) on their vehicle.

Ironically, after the first lunar landing, the mission that most Americans remember was the one that failed to achieve any of its goals except that of basic
survival. At 9:08 P.M. on 13 April 1970 James Lovell, Fred Haise, and Jack
Swigert, the crew of Apollo 13, heard a "sharp bang" and felt a wave of vibration move through the spacecraft. They were almost fifty-six hours into
their mission and some 205,500 miles (330,855 kilometers) from home. A
defective heater switch had caused the explosion of one of the two oxygen
ranks housed in the service module and a steady leak in the other. As the three
astronauts discussed the situation with Houston, their oxygen supply was
leaking away into space. (Amateur astronomers claimed to be able to see an
expanding cloud of gas trailing the CSM.) The fuel cells that generated electricity for the spacecraft were no longer operating, and there was no power to
gimbal the engine that would enable them to make an immediate return to
Earth.

The situation was critical. A lunar landing was out of the question. The
scramble was on to devise a way for the crew to survive. Some time before,
Grumman engineers had considered a situation in which the LM might have
to serve as a lifeboat for the crew. That time had arrived. Oxygen, water, and
power systems designed to keep two crewmen alive on the lunar surface
would be just sufficient to sustain all three astronauts as they swung around
the backside of the Moon and returned to Earth.

The three astronauts spent the next four days living, as one of them remarked, "like frogs in a frozen pond."[12] The temperature inside the command module dipped to 38 degrees F (3 degrees C). Moisture condensed on
the walls. They were chilled to the bone, with no way to prepare hot food.
During the long trip home, the dehydrated and exhausted men lost a total of
31.5 pounds (14 kilograms). An infection would keep Haise under a doctor's
care for three weeks after their return.

Back on Earth the crisis in space dominated the news. As a *New York Times*
editorialist noted, "Only in a formal sense will Apollo 13 go into history as a
failure." Throughout the period, churches remained open so people could
pray for the safe return of the crew of Apollo 13. The splashdown, four miles
(six kilometers) from the recovery ship *Iwo Jima* on 17 April 1970, capped
one of the epic stories of the space age with a happy ending.

Wernher von Braun left Huntsville in 1970 to become deputy associate administrator for planning at NASA headquarters in Washington, D.C. It was
not a pleasant time for him. Divorced from firsthand supervision of rocket
programs, worried about shrinking budgets for space, and finding himself

frozen out of serious decision making, he retired from the agency in 1973 to accept a post as vice president for engineering and development with Fairchild Industries in Germantown, Maryland.

At Fairchild, von Braun worked on behalf of the firm's ATS-F communications satellite effort. He traveled widely and remained a leading spokesperson for space flight. His lifelong good health already had begun to fail him, however. Battling cancer, he retired from Fairchild on 31 December 1976. Early the next year President Gerald Ford awarded him the National Medal of Science for a lifetime of contributions. Too ill to attend the ceremony, von Braun died in Alexandria, Virginia, on 16 June 1977 at the age of sixty-five. His death marked the end of an era.

People would remember Apollo 8, 11, and 13, with those missions' moments of triumph and crisis. The other Apollo crews did their best to establish an individual character for their missions. The astronauts of Apollo 12 identified themselves as the all-navy crew and played a scratchy rendition of "Anchors Away" during the course of their mission. The all–air force crew of Apollo 15 countered with "Off We Go into the Wild Blue Yonder."

For the first time since Gemini 3, beginning in 1969, the crews were allowed to select names for their spacecraft. Following the controversial loss of his Mercury capsule, Gus Grissom insisted on naming his Gemini 3 spacecraft the Molly Brown in honor of the unsinkable heroine celebrated in the Broadway musical. When NASA officials suggested that the name might not be appropriate, Grissom let it be known that his second choice was Titanic. Given that option, Gemini planners approved Molly Brown, but they insisted that subsequent spacecraft in the series would not be individually named.

The need to identify the two distinct Apollo spacecraft—the CSM and the LM—with call signs when separated resulted in a return to the earlier practice. The Apollo crews once again welcomed the opportunity to personalize and characterize their missions by naming their machines: Gumdrop and Spider (Apollo 9), Charlie Brown and Snoopy (Apollo 10), Columbia and Eagle (Apollo 11), Yankee Clipper and Intrepid (Apollo 12), Odyssey and Aquarius (Apollo 13), Kitty Hawk and Antares (Apollo 14), Endeavor and Falcon (Apollo 15), Casper and Orion (Apollo 16), and America and Challenger (Apollo 17).

Ultimately the attempt to create an individual character for each Apollo mission was a failure. Public interest and enthusiasm seemed to evaporate after the first Moon landing and the safe return of Apollo 13. A quarter of a cen-

The Apollo 15 spacecraft splashes down in the Pacific Ocean on 7 August 1971. (Courtesy of NASA)

tury later, our memories of the final Apollo missions are compressed, faded, blurred, and hopelessly entangled with other startling images from the nightly television news. The anonymous, space-suited figures hopping across the surface of another world, their faces masked by golden visors, are inextricably linked to scenes of a Vietnamese policeman blowing out the brains of a man whose arms are bound; of draft cards, bras, and entire neighborhoods going up in flames; of an anguished young woman kneeling beside a dead student on the campus of Kent State University in Ohio; of places like Haight-Ashbury and Woodstock. Ironically, or appropriately, the Apollo era has blended with the Age of Aquarius in our collective memory.

The Apollo voyages to the Moon were complete by the end of 1972. The value of the program has been endlessly debated since that time and will continue to be argued in the future. Was the crash program to put men on the Moon worth the enormous expenditure of time, money, and effort? Would a slower, more cautious, long-range manned space program with more modest goals have provided a greater scientific return or laid a firmer foundation for future space ventures? There are no simple answers to those and other questions.

We can be certain of one thing, however. The men and women of NASA accomplished the goal set for them eleven years earlier. People had journeyed to the Moon, explored its surface, and returned safely to Earth. Whatever its practical benefits, Project Apollo was a triumph of human ingenuity, will, and spirit. There was no mistaking the ultimate message of the program. If we could fly to the Moon, was there anything we could not accomplish?

14 • Salyut and Skylab
Rooms with a View

On 24 October 1969 Mstislav V. Keldysh, president of the Soviet Academy of Sciences, announced, "We no longer have a timetable for manned moon trips." It was a difficult admission for the unofficial chief scientist of the nation that had gloried in the triumphs of Sputnik and Yuri Gagarin. Nevertheless, the Soviets had no intention of abandoning their space program. They were prepared instead to channel their efforts in a new and interesting direction. Over the next three years, as six more Apollo crews journeyed to the Moon, cosmonauts would become the masters of near Earth space.

Soyuz would never fly to the Moon, but it finally provided Soviet mission planners with the capability of maneuvering in orbit. Soyuz 4 (Vladimir Shatalov) and Soyuz 5 (Boris Volynov, Alexei Yeliseyev, and Yevgeny Khrunov) completed the first rendezvous and docking of two manned Soviet spacecraft on 16 January 1969. Khrunov and Yeliseyev transferred to Soyuz 4, after which both spacecraft returned to Earth. In October 1969 Soyuz 6 (Georgi Shonin and Valery Kubasov), Soyuz 7 (Anatoli Filipchenko, Vladislav Volkov, and Viktor Gorbatko), and Soyuz 8 (Shatalov and Yeliseyev) conducted joint maneuvers in orbit. On 1 June 1970 Andrian Nikolayev and Vitaly Sevastyanov, the crew of Soyuz 9, remained in orbit for eighteen days, breaking the previous record for time in orbit held by Gemini 7. All of these flights were in preparation for the launch of Salyut 1 on 19 April 1971.

Salyut was the first true space station, a vehicle designed to serve the needs of a series of crews that would live and work aboard the spacecraft for extended periods of time. Measuring 47.6 feet (14.5 meters) in length (75 feet or 23 meters with a Soyuz attached to its docking port), Salyut consisted of a transfer compartment through which cosmonauts entered the spacecraft and three additional compartments in which they would eat, sleep, work, and relax. The vehicle was permanently fitted with a variety of telescopes, cameras, and scientific instruments. It also featured a rocket engine, burning unsymmetrical dimethyl hydrazine and nitric acid, which would lift the station to a higher altitude when its orbit began to decay. In addition several small thrusters were provided to orient the spacecraft. Initially, nine cosmonauts were trained for missions aboard Salyut 1. The first crew, Soyuz 10 (Shatalov, Yeliseyev, and Rukavishnikov), docked with Salyut on 24 April 1971 but were unable to enter the station, apparently because of a problem in equalizing the pressure between the two spacecraft.

Vladislav Volkov, Georgi Dobrovolsky, and Viktor Patsayev, the cosmonauts of Soyuz 11, launched on 6 June 1971, were the first to enter a Salyut in orbit. They remained aboard for three weeks, testing the various spacecraft systems and conducting scientific experiments. Reactivation of the Soyuz, its undocking, and its reentry seemed normal. However, when the recovery crew reached the descent module, they found all three crew members dead. A valve had ruptured during separation from the orbital module, allowing the air to leak out of the spacecraft. With three men in the module, there was no room for them to wear the spacesuits that might have saved their lives. The three cosmonauts were buried in the Kremlin wall, already the final resting place of Yuri Gagarin and Vladimir Komarov.

A redesign of the Soyuz spacecraft kept cosmonauts on the ground for almost two years. The orbit of Salyut 1 was allowed to decay, and it reentered the atmosphere and was destroyed on 11 October 1971, having hosted only one crew. NASA had a window of opportunity during which it would have near Earth space all to itself.

Skylab, the first U.S. space station, grew out of a series of discussions between officials at NASA headquarters and engineers at Huntsville and Houston during the early 1960s. The project was conceived as a fall-back program to be pursued if the Apollo lunar missions faltered. With the Moon landings a success, Skylab was even more appealing. It met the need for a relatively inexpensive but useful and impressive follow-on program to Apollo—a manned orbiting station based almost entirely on spare Apollo hardware.

The final plan called for a series of astronaut crews to take up residence inside the specially modified, Earth-orbiting third (S-4B) stage of a Saturn 5 rocket. Almost 59 feet (18 meters) long and 22 feet (6.7 meters) in diameter, the interior of the S-4B would be transformed into a comfortable home complete with three small bedrooms, a kitchen/dining area, a bathroom featuring a toilet and shower designed for use in zero gravity, an exercise area, and specialized work spaces. Skylab would be equipped with a medical facility, a photo lab, a broadcasting studio, and a sophisticated astronomical observatory. Enormous solar panels attached to the exterior of the S-4B would generate more than enough electricity to power the station.

The refurbished S-4B would be launched into a stable Earth orbit as the unpowered third stage of a Saturn 5. The crew, launched separately aboard a smaller Saturn 1B, would dock their standard Apollo command and service module to the upper end of Skylab, enter the station, and settle down for a long-term stay in space. Like Salyut, Skylab would support long duration orbital missions and enable astronauts to conduct a wide variety of important experiments in fields ranging from astrophysics to the production of new materials in space. There would be opportunities to investigate the medical consequences of long stays in weightlessness and to test new processes that might lead to space-based industries, and there would be time to conduct a full program of Earth-observation photography.

The last Saturn 5 to fly carried Skylab into orbit on 14 May 1973. The program almost came to an end before it began, when, during ascent, excessive vibration tore loose a protective micrometeoroid and sun shield, which in turn severed one of the two huge solar-cell arrays folded against the S-4B stage. Unable to deploy the second solar panel, which was pinned in place by a piece of the broken shield, ground controllers maneuvered a smaller solar panel designed to power scientific instruments into position facing the Sun. Telemetry indicated that the spacecraft now was receiving some power, but the temperature inside the cabin was rising to dangerous levels without the shield's protection.

The first Skylab crew—Charles Conrad Jr., Paul Weitz, and Joseph Kerwin—launched on 25 May 1973. Uncertain about what the astronauts would find when they reached the disabled spacecraft, NASA engineers equipped them with an assortment of tools and a hastily devised substitute sunshade. Over the next two days, operating out of the cramped Apollo CSM and the sweltering interior of Skylab, the crew succeeded in deploying a new shield and freeing the undamaged solar panel. Skylab was saved.

Three astronaut crews, nine men altogether, would live and work in Skylab for a total of 171 days between 25 May 1973 and 8 February 1974. They conducted nearly three hundred scientific, medical, and engineering experiments and overcame a host of problems. The astronauts found that it took time to become accustomed to weightlessness. They suffered bouts of illness in orbit, complained about bland food, and voiced resentment over the full work schedules prepared by mission planners. Complaints about overwork from the third crew led to a more reasonable schedule, complete with additional time for exercise and even a bit of relaxation.

In spite of the difficulties, the Skylab program represented NASA at its best. At minimum expense, the agency was able to operate a new and extraordinarily useful project that returned valuable information to science, medicine, and industry. To a much greater extent than Apollo, Skylab demonstrated that human beings could live and perform useful work in space.

The program came to an end all too soon. Three missions had depleted the stock of food and other consumables aboard the space station. Moreover, Skylab's orbit was decaying faster than predicted. Even at an altitude of 269 miles (433 kilometers), the spacecraft was fighting a losing battle with air resistance. Traces of the atmosphere gradually slowed its speed, dropping it ever closer to inevitable destruction in a fiery reentry. The end came on 11 July 1979. After six years and 34,981 orbits of the earth, Skylab plunged back into the atmosphere and disintegrated. Most of the chunks of the spacecraft that survived entry into the atmosphere fell into the Indian Ocean; a few small bits and pieces were later found in the outback of Western Australia.

NASA officials would put their surplus Apollo hardware to work one more time. The Apollo-Soyuz Test Project was the result of talks between President Richard M. Nixon and Soviet Premier Aleksey Kosygin. In view of the important role that space had played as a competitive arena during the 1960s, the two world leaders decided to underscore the new spirit of détente by undertaking a joint space mission.

The basic idea was simple enough: an Apollo CSM and a two-man Soyuz spacecraft would rendezvous and dock in orbit. It would not be easy. The U.S. and Soviet spacecraft featured incompatible docking mechanisms and life-support systems that operated with a different mix of atmospheric gases at different pressures. This incompatibility was the best of all possible reasons for flying the mission, however. ASTP would not only demonstrate the determination of the United States and the Soviet Union to cooperate but also bring American engineers and flight crews together with their Soviet coun-

The orbiting Skylab photographed from the Apollo command and service module after repairs to its sun shield and solar panel, 1973. The space station was manned for 171 days and helped demonstrate that humans could live and work productively in space. (Courtesy of NASA)

terparts, promoting technical exchange and preparing the way for future joint missions or international space-rescue efforts.

Slowly but surely, over a period of two years, U.S. and Soviet teams worked out the difficulties. ASTP astronauts Thomas P. Stafford, Vance D. Brand, and Donald K. "Deke" Slayton launched from the Kennedy Space Center on 15 July 1975, seven hours after cosmonauts Alexei Leonov and Valery Kubasov flew into orbit from a Soviet launch complex. Forty-five hours later U.S. mission commander Stafford nosed the new docking adapter up to the Soyuz spacecraft and locked it into place. The five men greeted one another in Russian and English, exchanged gifts, and began two days of joint scientific experiments. The Soyuz crew returned to Earth forty-three hours after separation. The Apollo astronauts brought the nearly flawless mission to a close with a safe splashdown on 24 July.

For the United States the tables were now turned. The crew of ASTP were the only Americans to fly into space between the conclusion of the Skylab program in February 1974 and the first flight of the Space Shuttle in April 1981. For NASA it was the end of an era. Never again would the roar of a Saturn rocket sweep across Cape Canaveral. Never again would an American astronaut ride a ballistic capsule into space. The few surviving examples of the giant rockets that had blasted the first human beings to the Moon and of the spacecraft in which the astronauts had made the journey were now museum pieces. The next time NASA sent an American into orbit it would be aboard a winged spacecraft. For seven years the cosmonauts would have space all to themselves.

The Soviets were preparing to resume operations in Earth orbit just as the U.S. Skylab program was coming to an end. They had accomplished a great deal during the two-year stand-down after the Soyuz 11 disaster. Cosmonauts

Soviet cosmonaut Alexei Leonov and American astronaut Donald K. "Deke" Slayton exchange points of view aboard the docked Apollo-Soyez spacecraft. The 1975 project brought the two countries together in one of the first cooperative space ventures, a product of the new spirit of détente. (Courtesy of NASA)

now would travel back into orbit aboard a redesigned and much improved Soyuz ferry spacecraft.

In appearance the new vehicle was much like the original Soyuz, although it lacked the solar panels that had made its predecessor self-sustaining in orbit. More important, the new spacecraft had seats for only two cosmonauts in spacesuits; the additional space was filled with life-support equipment. This redesigned Soyuz spacecraft would remain the workhorse vehicle of the Soviet manned space effort for the next nine years, ferrying crews to and from the orbiting Salyut space stations.

The Soviets launched a total of eight Salyut stations between 19 April 1971 and 19 April 1982. Salyut 2-1 failed to reach orbit, and Salyut 2 developed problems while in orbit and was never occupied. Of the remaining six, four (Salyuts 1, 4, 6, 7) were outfitted for civilian crews, and two (Salyuts 3 and 5) hosted missions of a primarily military nature. Salyuts 1 to 5 were first-generation spacecraft. Virtually all the supplies required by the visiting crews were packed aboard prior to launch. Any resupply would have to be ferried up with subsequent crews. Moreover, orbital decay was a problem. A standard Salyut burned 4.75 tons (4.32 metric tons) of propellant a year to maintain an altitude of 155 miles (250 kilometers). As a result Salyut 3 remained in orbit for less than a year (22 June 1974–24 January 1975), and Salyut 5 did only slightly better (22 June 1976–8 August 1977). Salyut 4 (26 December 1974–2 February 1977) was the most successful of the early stations. It hosted two manned missions: Soyuz 17 (10 January–9 February 1975) and Soyuz 18 (28 May–26 July 1975). In addition Soyuz 20 (17 November 1975–16 February 1976), an unmanned spacecraft carrying a variety of biological specimens, automatically docked with the station.

The Soyuz 18 cosmonauts, Pytor Klimuk and Vitaly Sevastyanov, spent a difficult sixty-three days aboard Salyut 4. Like the earlier Skylab crews, they were disturbed by nitpicking instructions from ground control. The spacecraft was cold, damp, and occasionally subject to vile smells. The green plants on which the cosmonauts lavished their attention withered and died within a few weeks. Reports reaching the West suggested that by the end of the mission the walls and windows of the Salyut were covered with mold.

Salyut 6 (29 September 1977–29 July 1982) and Salyut 7 (19 April 1982–6 February 1991) were second-generation vehicles capable of supporting missions of up to six months in duration. The new stations also featured two Soyuz docking ports. In addition the Soviets introduced the unmanned

Progress tanker spacecraft to ferry the extra propellant that was required to periodically lift the station back to a higher orbit.

A new Soyuz-T ferry craft entered service at the end of 1979. The vehicle was considerably larger than its predecessor, with a longer docking probe, a completely new propulsion system, reinstalled solar panels, and room for three space-suited cosmonauts.

Yuri Romanenko and Georgi Grechko, the crew of Soyuz 26, were the first to occupy the new Salyut 6 station and offered a view of things to come. Launched on 10 December 1977, they were joined by Vladimir Dzhanibekov and Oleg Makarov (Soyuz 27), who docked with the Salyut on 12 January 1978. The Soyuz 27 crew returned to Earth in the Soyuz 26 spacecraft on 16 January. Romanenko and Grechko remained aboard Salyut 6 during the first linkup with and refueling from Progress 1, the first of the new automated tankers.

On 2 March 1978 the cosmonauts welcomed the crew of Soyuz 28 aboard Salyut 6. Alexei Gubarev and Vladimir Remek, a Czech and the first non-Soviet "guest cosmonaut," remained on board until 10 March. Romanenko and Grechko returned to Earth aboard the Soyuz 27 spacecraft on 16 March 1978. They had spent ninety-six full days in orbit, breaking the previous record of eighty-four days held by the crew of Skylab 4.

The new record would be short-lived. The Soyuz 29 crew, Vladimir Kovalyonok and Alexander Ivanchenkov, spent 140 days aboard Salyut 6 (15 June–2 November 1978) and hosted two groups of visiting cosmonauts. Vladimir Lyakhov and Valery Ryumin of Soyuz 32 followed them aboard the station and remained for almost six months (25 February–19 August 1979) without visitors. Valery Ryumin and Leonid Popov of Soyuz 35 set the all-time record for Salyut 6: 185 days (9 April–11 October 1980). Having hosted a total of eighteen crews, the historic space station was abandoned. It reentered the atmosphere on July 29, 1982.

The Salyut 7 space station, launched on 19 April 1982, enabled the cosmonauts to remain aloft for even longer periods. Anatoly Berezovoi and Valentin Lebedev (Soyuz-T 5) inaugurated the new station with a record stay of 211 days (13 May–10 December 1982). It was a difficult seven months. The two cosmonauts became involved in an argument early in the mission and scarcely spoke to one another for the last four months aboard.

By the end of 1984 eight Soyuz crews (T 5 to T 12) had set up housekeeping aboard Salyut 7. Then, in February 1985, ground controllers suddenly lost contact with the spacecraft. Clearly, something was very wrong. Vladimir Dzhanibekov and Viktor Savinykh, the crew of rescue mission Soyuz-T 13 (6 June–25 September 1985), found the craft tumbling out of control. Gingerly docking with the Salyut, they were able to reestablish attitude control and to reenter the dead, cold spacecraft.

Working out of the docked Soyuz-T 13, they gradually brought the Salyut back to life. Vladimir Vasyutin, Georgi Grechko, and Alexander Volkov, the crew of Soyuz-T 14, joined them in orbit on 18 September and remained aboard Salyut 7 until 21 November. The Soyuz-T 15 crew (13 March–5 May 1986) paid a final visit to Salyut 7, retrieving scientific experiments left by their predecessors and mothballing the station.

The Soviets launched a total of forty-three manned missions between 1971 and 1985. One of those flights was aborted before reaching orbit; five were unable to dock with the Salyut; three, including the ASTP mission (Soyuz 19), did not involve a visit to a space station. Thirty-five Soviet cosmonauts lived and worked aboard the Salyuts. In addition ten Intercosmos cosmonauts, representing France, India, Mongolia, Cuba, East Germany, Hungary, Poland, Romania, Czechoslovakia, and Vietnam, spent time aboard the Salyuts as guests of the Soviets.

Only one of the forty-five Salyut residents was a woman. Svetlana Savitskaya was the second Soviet woman to venture into space, the first woman to make two flights (Soyuz-T 7 and T 12), and the first woman to walk in space. Vladimir Dzhanibekov and Gennady Strekalov flew four missions apiece. Ten cosmonauts flew three missions; fifteen served on two crews. During the entire period, the cosmonauts logged more than 3,400 man-days in orbit. In that regard Valery Ryumin held the personal a record, logging 358 days in space during three missions aboard the Salyut 6. "I was afraid most of an appendicitis attack," he later recalled. "Also, I was afraid of getting a toothache. Once I dreamed I had a toothache. I awoke almost instantly, feeling, yes, my tooth really did hurt. But, by morning, the pain was gone."[1]

Ryumin was hesitant about space walking as well. "Speaking frankly, I was not exactly eager to step out into the void," he explained. "It was scary."[2] Most cosmonauts would return to Earth with their own memories of a frightening moment or two. Alexander Alexandrov and Vladimir Lyakhov, the crew of Soyuz-T 9, were in the middle of a discussion with ground controllers when a loud crack reverberated through Salyut 7. A micrometeoroid had left a dark streak across one of the cabin windows.

The long missions were especially difficult for family men. Leonid Kizim's daughter was born three months into a mission. For the next six months she was nothing more to him than a picture on the cabin television. Mail and videotapes that came up with visiting crews were always especially welcome. Anatoly Berezovoi played a videotape of his daughter's birthday party when he was feeling homesick. "You watch, get engrossed in it, and it seems like you are with your family."[3]

Long-duration space flight literally changed the cosmonauts' and astronauts' bodies. Skylab and Salyut crews grew as much as an inch in height after weeks in microgravity. "At last," Pete Conrad quipped, "I'm taller than my wife."[4] Fluids tended to gather in the center of the body, so that although legs became noticeably thinner, chests expanded and faces took on a bloated look. Those alterations, combined with what happens to hair in a weightless environment, could make a person all but unrecognizable.

Exercise equipment aboard Skylab was designed to both gather medical data and help the astronauts maintain physical condition while living in a microgravity environment. The bicycle ergometer was never a particular favorite. In order to operate the machine an astronaut had to be literally belted, cinched, and buckled into place. "I was really running out of gas," Pete Conrad complained after a few minutes on the ergometer. "And yet I was using muscles that I don't normally use on the ground."[5]

Creature comforts were few and far between aboard both Skylab and Salyut. Take something as straightforward as water quality, for example. The air used to pressurize the water supply aboard Skylab produced bubbles that did not rise to the surface in microgravity. Occasionally, escaping bubbles would burst rehydrated food bags, filling the cabin with globs of floating food. The air bubbles also contributed to flatulence. As a member of the final Skylab crew noted, "Farting about 500 times a day is not a way to go."[6]

For their part the cosmonauts were treated to instant coffee made with water reclaimed from cabin humidity and waste water. Passage through filters removed the contaminants, but not the thought that one was drinking processed perspiration and dishwater.

A great deal of that perspiration was generated by the rigorous exercise required to tone muscles that no longer had to fight the pull of gravity. An orbital workout was often a less-than-pleasant experience. In the absence of gravity, perspiration accumulates near the spot where it is generated. Skylab astronauts complained that a pool of sweat would begin to build above the breastbone and grow larger as exercise continued. "You get warm quickly," one cosmonaut explained. "The face and arms are sweaty, and by the end of training you are soaked."[7]

Normal washing was done with towels and premoistened wipes. Cosmonauts averaged a shower once every two weeks. As Alexander Ivanchenkov described the process, that was quite enough:

We turn on the electric water heater and let down a polyethylene cylinder [the shower] from the disc [on the ceiling] to the floor. Fastened to the bottom of this chamber are rubber slippers for the feet, to prevent you from floating upwards. We then get undressed, unzip the cylinder's watertight zipper, climb into the shower and seal ourselves in the same manner. Above our head [are] the showerhead and cellophane bags containing napkins and a towel. Before turning on the water we place in our mouth a mouthpiece leading outside the chamber and we block our noses using clamps. Next we open the bag containing a small towel that has been saturated with soap and turn on the water.[8]

The Skylab shower was not much better. In both cases the worst part was cleaning up after. The first Skylab crew showered once a week. Subsequent occupants made do with a daily once over with a wash cloth.

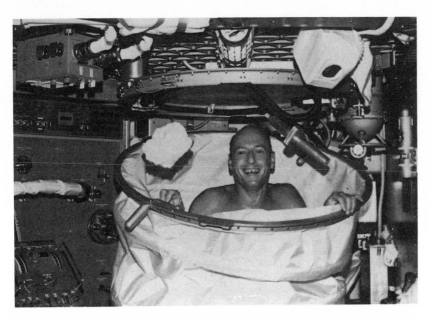

Astronaut Charles Conrad Jr. demonstrates the shower aboard Skylab 2, 1973. Despite Conrad's apparent satisfaction with the apparatus, most crew members opted for a daily once over with a wash cloth. (Courtesy of NASA)

The most frequently asked question of any astronaut or cosmonaut was, "How do you go to the bathroom in space?" As one NASA official source noted, what was euphemistically known as the Skylab "waste-management compartment" in fact "resembled the bathroom of a commercial jet liner in its size, metallic appearance, and even its gurgling noises." There were problems, however. One astronaut pointed out that, for ease of cleaning, the triangular gridwork common to the rest of the spacecraft (into which an astronaut would lock the soles of his shoes) was not employed in the bathroom. As a result, an incautious space voyager might "just ricochet off the wall like a BB in a tin can" at a most inopportune moment.[9]

Food was a problem for both cosmonauts and astronauts. "I found that if I reconstituted the peas, the beans and the asparagus early, and then reheated them," Pete Conrad quipped, "I still didn't like them, but they were a lot easier to choke down than when I added the hot water, shook up the bag and then tried to get them down."[10]

The psychological reaction to long duration stays in space varied with the individual. Some thrived on it. Others, including Leonid Popov and Alexander Ivanchenkov, battled deep depression. Anatoly Berezovoi and Valentin Lebedev extended their sleep time by two hours in an attempt to deal with tension. A touch of humor was a particularly important factor in maintaining psychological balance. Early in the course of one two-man mission, a cosmonaut settled down for a nap in the docked Soyuz spacecraft. His worried companion, unable to find him in the Salyut, knocked on the closed hatch of the Soyuz. Startled awake, the dozing cosmonaut asked, "Who's there?" They were still laughing at the end of their mission.[11]

Salyut cosmonaut or Skylab astronaut, first-generation space dwellers never seemed to tire of looking out the window. Skylab had almost flown without a window. Members of the Huntsville team responsible for the design of the vehicle pointed out that although a window would be a nice addition to the spacecraft, it would add to the expense of the project and reduce the perfect structural integrity of the original S-4B stage. Concerned about the problem, NASA official George Mueller invited the famous industrial designer Raymond Loewy to attend a meeting with the Huntsville team in the fall of 1969. Surprised that the question was even under discussion, Loewy remarked that it was unthinkable to send human beings into orbit and not provide them with a window to enjoy the view. "And thus," as astronaut Mike Collins explained, "the issue was decided."[12]

As cosmonaut Anatoly Berezovoi noted: "It is important to understand that for us in orbit, visual observation replaced theater and movies."[13] Shuttle astronaut Joe Allen points out that, from orbit, "you see things in three dimensions. You would think that from orbit, the globe would look fairly flat, as though you were in a high flying airplane. But, in fact, that is not the case. . . . You can see the mountains coming up from the surface of the earth. You can see the clouds well above the surface of the earth. You can see the shadow of the clouds on the earth. That shadow, that relief of clouds and mountains and earth, is very striking and gives you a sharp feeling of three dimensions."[14]

They watched clouds, observed changes in color as the seasons moved across the face of the land, and saw lightning bursting over the dark surface of the earth. Most of all, they watched the sea. One astronaut was fascinated by the fact that "each crewman saw sea-surface features that the others did not notice." Bill Pogue noticed the distinct colors of the silt associated with the various tributaries of the Amazon; others called attention to the patterns of turbulence created by the passage of ocean currents around islands. Plankton blooms, iceberg formation and movement, the precise delineation of currents, and a variety of other phenomena were sources of endless fascination.

"The Sea of Okhotsk is brownish green, and the Caspian is a deep, dark blue," reported one cosmonaut. "It [the ocean] is multicolored, iridescent, patterned, as if it were alive. We saw the track of a ship. Guadeloupe Island seemed to be swimming in the ocean."[15] Those are the words of a homesick man. It was a common phenomenon in space.

"While out there in orbit," Vladimir Lyakhov noted, "dreams were usually about the Earth."[16] Dispensers of pine scent were popular with the cosmonauts. Alexander Ivanchenkov kept replaying a cassette recording of "a rooster crowing, a mooing cow, flowing water," and other Earth sounds. "What a pleasure that was," he recalled.[17] Anatoly Berezovoi agreed: "We switched them on most frequently of all, and we never grew tired of them. They were like meetings with Earth."

Indeed, few of the first-generation dwellers in space seem to have enjoyed the experience. "All the inexperienced cosmonauts agitate for longer space flights," Oleg Gazenko noted, "but the veterans of long-term flights don't."[18] *Pravda* gave Valentin Lebedev's memoirs of a long stay in orbit a title that aptly reflected the contents: *His Heart Remains on Earth.* Gazenko, a physician who headed the Soviet space medicine effort, remarked that no one hated space flight more than Valery Ryumin, who, for a considerable period of time, held the record for the most time spent in orbit.

Perhaps Vladimir Lyakhov best summed up the matter in describing his own reaction, and that of his companion Ryumin, to a book of nature photographs sent up with a visiting crew: "It was pleasant to turn the pages and once again look at the running brooks, at the rivers and lakes, to remember the tie with Earth, from that which we had left, but where we will always strive to return. It is our home."[19]

15 • The Shuttle Era

The conquest of space sounded simple when Wernher von Braun described it in *Collier's* magazine and on Walt Disney's television show in the 1950s. It would be a step-by-step process, beginning with the launch of unmanned satellites, followed by short trips into space aboard winged rockets, and culminating in the construction of a permanent, Earth-orbiting space station that would serve as a base for later manned voyages to the Moon, to Mars, and beyond. It was a rational, well-conceived plan aimed at establishing a permanent human presence in space.

But Americans did not fly to the Moon in response to a sense of cosmic manifest destiny or because Wernher von Braun thought they should. Thirty years of experience with large ballistic weapons, from the era of the V-2 to that of the Atlas and the Titan, had brought the Moon within our reach. The pressures of international competition and the need to provide a stunning demonstration of American strength and technological prowess were reasons enough to stretch a bit and try to grasp the prize.

The Apollo voyages were a product of the Cold War—a deliberate and carefully crafted response to the Soviet challenge. Von Braun himself had quipped that the first U.S. astronauts to reach the Moon might have to pass through Russian customs. For the first decade of its history, the principal job of NASA was to ensure that American space travelers would not require a visa. Time was of the essence. In an attempt to journey to the Moon in the quickest, simplest, and most direct fashion, the agency leapfrogged von

Braun's systematic blueprint for space flight. Mercury, Gemini, and Apollo were designed to beat the Soviets to a lunar landing, not to be reusable or to provide the foundation for a long-term future in space.

NASA personnel achieved the major objective established by the nation's political leaders only to discover that they bad reached a technological dead end. The spacecraft that had transported astronauts to the lunar surface had no potential for development into later-generation vehicles capable of more distant journeys into space. Furthermore, given the fact that a Saturn 1B was the smallest launcher capable of boosting an Apollo command and service module into Earth orbit, it was not particularly economical.

Even before the first step on the Moon NASA scientists and engineers had begun to take a second look at Wernher von Braun's blueprint for space flight. It was impossible to overlook the fact that any post-Apollo planning for future manned space programs would have to include a reusable vehicle capable of traveling into orbit, remaining on station for limited periods of time, and then returning to Earth. As von Braun had suggested, such a craft would generate lift and fly back from space for a runway landing. The vehicle would become the cornerstone of all future programs, including a permanent space station.

U.S. Air Force interest in a manned, aerodynamic spacecraft that could fly home through the atmosphere dated to the early 1960s. A series of top-secret engineering studies and experiments had explored a number of options, ranging from a second-generation X-15 to plans for the X-20 Dyna-Soar. These projects were eventually shelved, however. By the mid-1960s the joint Department of Defense–NASA Aeronautics and Astronautics Coordinating Board had begun to search for common ground on which to unite the interests of military and civilian space planners.

On the basis of those talks, administrators at Marshall Space Flight Center and the Manned Spacecraft Center issued a call for the design of a fully reusable Integrated Launch and Reentry Vehicle (ILRV) capable of carrying a 40,000-pound (18,100-kilogram) payload into low Earth orbit. By April 1969 NASA had established the Space Shuttle Task Group, on which the air force was fully represented, and had issued contracts for ILRV design studies to North American Rockwell, Lockheed, McDonnell Douglas, and General Dynamics. These phase A studies led to the awarding of more detailed phase B contracts to North American Rockwell and McDonnell Douglas. Unwilling to limit themselves at this early stage, NASA officials also offered alternate Shuttle design-study contracts to Lockheed, Chrysler, and a Boeing/ Grumman consortium.

The time had come to obtain political support for NASA's plans. On 15 September 1969 the Space Task Group headed by Vice President Spiro Agnew presented President Richard Nixon with three options for the future of the U.S. space effort. For $5.5 to $8 billion a year over the next six years, NASA could send people to Mars, build a lunar base, establish an Earth-orbiting space station, and develop its winged, reusable spacecraft. The second option, costing less than $5 billion per year, included development of both the Space Shuttle and the Mars mission. Finally, a bargain-basement special would include only the Earth-orbiting station and the Space Shuttle. President Nixon chose the third and least expensive option, deferring space station development but leaving the door open for NASA to return with a proposal for additional funding to cover a no-frills manned program.

NASA administrator Thomas Paine identified the critical need for what already was being called the Space Shuttle. He described it as an all-purpose space truck. Each vehicle would make repeated trips into space, hauling satellites aloft, orbiting the Earth while the crew of astronauts and mission specialists conducted their experiments, and transporting materials with which to build a permanent space station. The basic vehicle would operate only in Earth orbit. A special "space tug," essentially a strap-on rocket engine, would boost lunar and interplanetary spacecraft that had been carried into orbit by the Shuttle on to their destinations.

Selling the Shuttle to the president and Congress after the expenditures of the Apollo era was a difficult task. Prior to his departure from the agency late in 1970, Paine had decided that the Shuttle could be sold only as the complete answer to the nation's total space requirements for the immediate future. There would be no more need for either NASA or the military services to purchase traditional throwaway launch vehicles. The cost savings would be enormous. Almost at once, however, NASA was forced to rethink its plans for a dream Shuttle. As a result of post-Apollo budget tightening in 1971, the Office of Management and Budget (OMB) announced that NASA's annual appropriation would be limited to $3.3 billion. Conservative estimates for the development of a fully reusable Shuttle then stood at $12 billion. NASA would have to fall back to a less expensive, partially reusable configuration.

The phase B industry studies produced designs that NASA could no longer afford. As a result, air force support for the program became even more crucial. Air force officials, aware that NASA would need their cooperation and support to obtain approval for the new spacecraft, insisted on some measure of design control before agreeing to book payload space aboard the proposed Shuttle. The craft would have to handle payloads of up to 40,000 pounds (18,100 kilograms) and feature a payload bay measuring at least 15 feet by 60 feet (5 meters by 18.3 meters) and a cross-range capability that would enable it to land as far as 1,500 miles (2,420 kilometers) off its orbital track. These requirements would have to be met if the proposed craft was to handle USAF reconnaissance satellites.

Paine's successors, Acting Administrator George Low and Administrator James C. Fletcher, were faced with the prospect of designing a vehicle that represented a compromise between USAF requirements and their budget limitations. The heyday of the Apollo blank check was over. In the winter of 1972, after fending off OMB's demands for yet another reduction in the scale of the program and the size of the vehicle, NASA received presidential approval to proceed with design studies for a craft with a price tag of $2.6 billion.

Based on the USAF requirements and the results of the two phases of industry studies, engineers at the Manned Spacecraft Center in Houston developed a baseline design of their own. The crew would travel to and from space in a delta-winged orbiter. In an effort to save weight, space, and money, the propellant for the main engines would be housed in an external tank that could be jettisoned when emptied. Two large, solid-fuel boosters would be strapped onto the external tank for additional thrust during the early portions of the flight. The Houston design for the Space Shuttle served as the basis for final bids. North American Rockwell was awarded the contract for the orbiter, Martin Marietta would produce the external tank, and Thiokol would construct the solid rocket boosters. Work was underway on all phases of the project by the spring of 1974.

The redesigned Shuttle would be smaller and more expensive to operate than the larger, fully reusable designs of the phase A and B studies, but it could be developed and built for roughly half the cost. Throughout the early 1970s, however, OMB officials insisted on additional cost reductions. Representative Olin "Tiger" Teague of Texas and other NASA supporters in Congress fought for the Shuttle program, but appropriations continued to fall below the amount that many engineers honestly believed was required to do the job. NASA officials saw little choice but to forge ahead with the money allocated.

The Space Shuttle Discovery. The Shuttle's final design called for a delta-winged orbiter with an external propellant tank that could be jettisoned after launch. Two large solid-fuel boosters would be strapped onto the tank for additional thrust. (Courtesy of NASA)

The agency had bet its entire future on the Shuttle. With the completion of Skylab and of Apollo-Soyuz Test Project programs, no more Americans would fly into space until the Shuttle was operational. Satellite and planetary-exploration programs were dependent on the new vehicle as well: there were no funds available for the development of more traditional launchers, and the existing stock of such rockets was dwindling.

Obstacles abounded. The Shuttle would be protected from the heat of reentry by roughly thirty thousand fragile, lightweight silica tiles glued to its skin. Traveling through the upper layers of the atmosphere at a speed of 18,000 miles (28,980 kilometers) per hour, the spacecraft would experience a skin temperature of some 3,000 degrees F (1,649 degrees C). Given the speeds, temperatures, and pressures to which they would be subjected, would the tiles hold up? Without them, the structure of the spacecraft would not survive.

In view of its mixed character as a true aerospace craft, the behavior of the

Shuttle at high reentry speeds also was cause for concern. Computers would be responsible for attitude control until the speed was reduced to the point where the pilot could begin to maneuver the craft. The slightest misalignment during reentry could result in a hypersonic spin that would destroy the spacecraft before the pilot could recover control.

Then there were the engines. The orbiter would sport a grand total of forty-nine rocket engines: three Space Shuttle main engines (SSMEs), forty-four reaction control system (RCS) engines, and two orbital maneuvering system (OMS) engines. Thirty-eight of the RCS engines, clustered at three points on the spacecraft, would be the primary RCS thrusters, which would develop 870 pounds of thrust (3,870 newtons) each. The remaining six RCS engines, called vernier engines, would produce 25 pounds of thrust (110 newtons). Both the RCS and OMS engines would burn monomethyl hydrazine (fuel) and nitrogen tetroxide (oxidizer) and would be used to control the attitude of the spacecraft in orbit. The OMS engines, mounted in pods at the rear of the spacecraft, would develop 6,000 pounds of thrust (26,700 newtons) apiece. They would boost the Shuttle into orbit, enable it to change orbits in space, and provide the retrofire for return to Earth at the end of a mission.

The real problems, however, would be encountered with the SSMEs. Housed together on the tail of the orbiter, they would burn liquid hydrogen and liquid oxygen from the external tank, producing 375,000 pounds of thrust (1,668,750 newtons) each. The SSMEs would not be the most powerful rocket engines ever built, but they surely would be among the most complex.

For most large rocket engines, success or failure is defined by performance on a single launch. The SSME program set out to develop engines capable of propelling as many as one hundred missions into space. The engines were also required to be throttled over a range from 60 to 109 percent of their rated thrust and to operate with combustion-chamber pressures more than three times as high as those of the F-1. The turbopumps would spin at thirty-five thousand revolutions per minute, and feed 1,000 gallons (3,846.2 liters) of propellant per second to the combustion chamber. Finally, the engines had to be very light.

A great many things could, and did, go wrong with these new and complex engines. Turbopumps disintegrated, propellant lines cracked and broke, and combustion chambers burned through. Time passed, the problems grew, and costs rose. A 1978 report by a committee of the National Academy of Sciences suggested that NASA had underestimated both the difficulties and the costs of developing the SSMEs. As they struggled to overcome the myriad serious technical problems plaguing the Shuttle, NASA officials convinced themselves that they were making do in the best tradition of the agency. As had been the case prior to the Apollo 204 fire, however, some critics believed that potentially dangerous flaws were being overlooked in the rush to build and fly a new spacecraft with ever shrinking resources.

Ultimately USAF leaders would be counted among the most enthusiastic proponents of the Space Shuttle. While paying a 1977 call at the Pentagon office of Dr. Hans Mark, deputy secretary of the air force, William Pickering, director of the JPL, was amused to see a large and nicely detailed model of the Shuttle, marked and painted as a USAF vehicle. At critical meetings with Carter administration officials in November and December 1977, Mark and Secretary of the Air Force Harold Brown provided strong support for the Shuttle, which proved decisive in preventing the planning effort from being reduced to an experimental program.

By early 1978 it looked as though NASA decision makers had won their bet on the Shuttle. The first vehicle, Orbiter 101, named the Enterprise in honor of the starship featured on the television series "Star Trek," was loaded on a ninety-wheel trailer at the Rockwell International plant in Palmdale, California, and trucked a short distance to NASA's Dryden Flight Research Center. In time there would be five more Shuttle orbiters, each named after a famous ship: Columbia, Challenger, Discovery, Atlantis, and, most recently, Endeavor.

When operational, the Shuttles would be launched from the Kennedy Space Center in Florida and would land at Edwards Air Force Base in California, adjacent to Dryden. Additional landing strips were prepared at White Sands, New Mexico, and Kennedy Space Center, Florida. Ultimately the USAF hoped to launch and land its own Shuttle missions from Vandenberg Air Force Base, California. For a time, however, the optimum mission called for a launch from the Cape and a landing at Edwards, after which the orbiter would be hoisted into place on top of a specially equipped Boeing 747, officially known as the Shuttle Carrier Aircraft (SCA), and flown back to the cape for its next voyage. The first series of five airborne test flights of the new vehicle were conducted with Enterprise riding piggyback on the SCA. On 12 August 1977 the crew broke away from the airborne 747 and glided the unpowered Enterprise back to a safe landing at Edwards. Four more flights of this sort completed the initial test program.

Columbia, the first Space Shuttle to fly into orbit, roared aloft from the cape on 12 April 1981 with astronauts John W. Young and Robert L. Crippen at the controls. The program was running more than two years late and $1 billion over budget. None of that seemed to matter. For the first time in six years U.S. astronauts had traveled into orbit. Three more test flights followed in quick succession: Space Transportation System (STS) 2, 3, and 4 (12 November 1981–27 June 1982). Each of the two-man STS crews tested spacecraft systems and explored the capabilities of the new vehicle. STS-3 (Jack Lousma and Gordon Fullerton) set the flight-duration record with an eight-day voyage and made the first landing on the runway at White Sands, New Mexico.

With the launch of STS-5 on 11 November 1982, NASA announced the entry of the Space Shuttle into operational service. The USAF and other potential users of the Shuttle still were not convinced, however. In February 1984 a national-security-decision directive suggested that NASA's STS could not be regarded as truly "operational" until it was conducting its promised twenty-four launches a year. NASA was never to achieve that goal. In the end the four orbiters made twenty-four trips into space between 21 April 1981 and 28 January 1986, when the Challenger lifted off on Shuttle mission S1-L. The program nevertheless accomplished a great deal. Ninety-six astronauts rode into space aboard the Shuttle during its first five years of operation. Sally Ride, one of eight female astronauts, was the first American woman to fly into space; Kathryn Sullivan was the first to make a space walk. All told, between 1981 and 1986, U.S. astronauts spent a total of 153 days, 3 hours, and 20 minutes in orbit.

The Space Shuttle program brought change to the astronaut corps. Prior to the beginning of operational Shuttle missions, virtually all astronauts were professional pilots. William Benjamin Lenoir and Joseph Percival Allen of STS-5 (11–16 November 1982) were the first of a new breed of astronaut, the mission specialists. Although some mission specialists were pilots, they were selected for their credentials as scientists or physicians and were responsible for the conduct of research during space missions.

Ulf Merbold and Bryon Lichtenberg of STS-9 (28 November–8 December 1983) added yet a third category of personnel to the roster of NASA flight crews, the payload specialists. Usually these were scientists or engineers selected to conduct specific experiments or tests in space, and they worked under the supervision of the mission specialists. From 1981 to 1986 eight foreign guests—three from West Germany and one each from Canada, Mexico, the Netherlands, France, and Saudi Arabia—flew as payload specialists. In

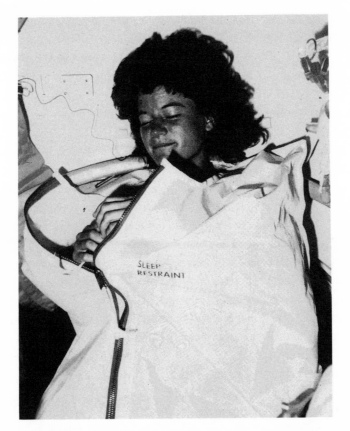

Astronaut Sally K. Ride, shown here in one of the Space Shuttle's sleep restraint devices, became America's first woman in space in June 1983. (Courtesy of NASA)

addition NASA launched three special passengers: Senator Edwin "Jake" Gain (51-D, 12–19 April 1985); Congressman William Nelson (61-C, 12–18 January 1986); and teacher Christa McAuliffe (51-L, 28 January 1986).

Beginning with 51-S (11–16 November 1982), mission and payload specialists conducted a variety of scientific experiments and tests designed to demonstrate the potential for industrial processing in space. Spacelab, a specially equipped, closed-habitat module carried in the Shuttle's cargo bay, was their domain. Financed by the European Space Agency (ESA), the first Spacelab was carried into orbit on the STS-9 mission (28 November–8 December 1983). Three other Spacelab missions were flown between 1983 and 1985.

NASA demonstrated that the Shuttle was capable of performing a wide variety of missions. By 1985 the new spacecraft had carried several dozen satellites into orbit. The crew of mission 41-C (3–11 April 1984) captured, repaired, and returned to orbit the malfunctioning Solar Max satellite. Discovery 51-A (8–16 November 1984) retrieved two malfunctioning communications satellites from orbit and brought them home for repair and relaunch.

In spite of these successes, NASA was unable to make good on its promise of an STS capable of meeting all of the nation's needs. Agency officials had forecasted a launch every two weeks once the STS hit its stride. In practice, engineers and technicians were forced to work overtime to reduce the interval between flights to two months. The orbiters required more extensive refurbishing following each flight than had been predicted, and the availability of spare parts was a constant problem.

The air force had never been comfortable with the notion of total reliance on NASA for launch services. During the 1980s reconnaissance satellites had become critical to national defense. Repeated launch delays and an uncertain schedule of Shuttle flights convinced USAF officials that they could not rely on NASA in an emergency. As a result, the Pentagon requested and received congressional approval for a new generation of traditional throwaway launch vehicles to meet USAF's specific needs.

NASA also was unable to make good on its promise of enormous savings in launch costs. Initial cost estimates were based on a full payload bay for every Shuttle mission and a heavy schedule of flights. In fact, NASA was forced to subsidize launch costs to keep its payload bays even partially full. The agency would have had to charge ten times its standard rate to send a pound of payload into orbit in order to cover the full cost of Shuttle development and operation with launch fees. At that price the orbiters would have flown with virtually empty payload bays. Even with subsidies, the leaders of the U.S. civilian space program had to scramble for launch fees. By the early 1980s NASA was facing stiff competition from the ESA.

The ESA was rooted in two predecessor organizations founded in 1964 by Great Britain and a handful of other European nations interested in developing a space-flight capability. The European Launcher Development Organization and the European Space Research Organization collapsed with the failure of their Europa 1 and 2 launch vehicles in 1971. ESA took their place and launched the first of its Ariane 1 rockets on 24 December 1979. Hungry for launch contracts and willing to tailor its services to meet the needs of commercial firms, scientific-research groups, or small nations, ESA set itself up in the space-launch business. In the spring of 1984 the organization conducted its first commercial launch, the Spacenet F-1 communications satellite, for the GTE Corporation. It was the first of many launch contracts that ESA would lure away from NASA and the Shuttle program.

Faced with continuing budget problems, NASA seized every opportunity to build both congressional and public support for an expanded civilian space program. It flew two of its leading congressional supporters into space, and it publicized inexpensive "getaway-special" launch rates for small, experimental packages in order to drum up additional business and focus attention on the program. The agency also held a nationwide student-science competition, flying the winning experiments into space at no charge. An announcement that NASA would fly a teacher aboard the Shuttle drew applications from eleven thousand educators across the nation.

The winner of the teacher-in-space competition, Christa McAuliffe of Concord, New Hampshire, was scheduled to fly aboard the Challenger on mission 51-L. McAuliffe and Gregory Jarvis, an engineer and satellite designer from Hughes Aircraft, were the payload specialists. Francis Richard Scobee, a veteran of one previous flight, was the mission commander, Michael Smith was the pilot, and Ronald McNair, Ellison Onizuka, and Judith Resnik were the mission specialists, responsible for the scientific and experimental work conducted in orbit.

The payload for the planned six-day flight included a NASA communications satellite, a tracking-and-data-relay satellite with an inertial upper-stage booster attached (TDRS-B/IUS), and a small Spartan spacecraft developed by NASA and the University of Colorado to study Halley's comet. There were also plans to monitor the comet from the orbiter throughout the mission. In addition, the mission specialists would conduct four major experiments, including one of the winning entries in NASA's program for high school science students.

Challenger lifted off the pad late on the morning of 28 January 1986. It was NASA's veteran orbiter, having carried a total of fifty-four crew members into space during the course of nine previous missions. One minute and thirteen seconds into its tenth flight, Challenger vanished in a cloud of smoke and flame. Gus Grissom, Ed White, and Roger Chaffee were no longer the only casualties of the U.S. space effort

The next thirty-three months were the most difficult period in NASA's history. In the immediate aftermath of the tragedy President Ronald Reagan appointed William P. Rogers, former attorney general and secretary of state, to head a distinguished commission charged with investigating the accident and recommending a future course of action. The members of the Rogers Commission—Neil Armstrong, Chuck Yeager, Sally Ride, and Richard Feynman—worked quickly and delivered their final report in four months.

The commission had little difficulty identifying the specific cause of the disaster. The rubber O-rings sealing the joint between the bottom two segments of the right-hand solid-propellant booster had failed, allowing flames to burn through the outer casing of the booster and then through the skin of the external liquid-propellant tank. But the problems went much deeper than the failure of a single part of the spacecraft. Engineers at Morton Thiokol, the firm that built the solid boosters, had expressed serious concern about the performance of the pressure seals. A study of seals recovered from previous flights had revealed several instances of damaged O-rings. The worst examples came from Shuttles launched on days when it had been cold at the cape.

The day before the Challenger flight, as it became apparent that this would be the first Shuttle launch conducted in below-freezing temperatures, a series of telephone conferences took place between nervous Thiokol engineers in Utah and NASA specialists at Huntsville. At one point Thiokol's vice president of engineering recommended that the Challenger launch be postponed until the temperature rose to 53 degrees F (12 degrees C), the previous record low for a launch. One NASA official responded by asking if the Thiokol experts really expected the agency to hold off the launch until spring. Thiokol managers, anxious to please their most important customer, called back late that night and withdrew their objections.

At NASA the desire to keep the program moving had overwhelmed considerations of safety. Agency officials had dismantled the intricate network of quality-assurance procedures instituted by James Webb, returning primary responsibility for safety to program managers and the heads of the individual research centers. No one in authority at NASA headquarters, Houston, or at the cape had been seriously involved in the afternoon and evening debates about the seals on the solid boosters. Officials at Huntsville, the center responsible for supervising work on the boosters, made the decision on their own.

The members of the Rogers Commission not only pinpointed the cause of

The Challenger disaster on 28 January 1986 began the most difficult period in NASA's history. (Courtesy of NASA)

the accident but also provided a list of other potentially catastrophic technical problems. They made nine basic recommendations, ranging from the obvious suggestion that the solid-rocket booster joints be redesigned to more far-ranging proposals relating to the management of the agency. The commission recommended that final responsibility for safety be placed in the hands of an independent group reporting directly to the NASA administrator, as had been the case under James Webb. In addition, members suggested that NASA managers improve communications between the centers and headquarters, undertake an immediate and thorough safety review of all Shuttle systems, address the problem of landing safety, and develop an improved crew-escape system. Finally, the Rogers Commission advised that NASA establish a slower and more reasonable launch schedule and take whatever steps were necessary to ease the "relentless pressure" of operating the nation's only space vehicle.

The report was blunt and honest, identifying major problems that had been growing throughout the post-Apollo years. At the same time the members of the commission were careful to express their confidence in the ability of the agency to correct its problems, urging that "NASA continue to receive the support of the Administration and the nation":

The agency constitutes a national resource that plays a critical role in space exploration and development. It also provides a symbol of national pride and technological leadership. The Commission applauds NASA's spectacular achievements of the past and anticipates impressive achievements to come. The findings and recommendations presented in this report are intended to contribute to the future NASA successes that the nation both expects and requires as the 21st century approaches.[1]

The Apollo 204 fire killing astronauts Grissom, White, and Chaffee had occurred at a time when NASA was racing to achieve a monumental goal and morale within the agency was at its peak. The need to reorganize and push forward toward the primary objective of a lunar landing helped pull NASA through that crisis. The Challenger disaster, however, struck an agency already hard pressed by the constant demand to meet impossible schedules and deadlines. Morale plummeted in the wake of the tragedy. James Beggs, the administrator of NASA, had resigned two months before the Challenger explosion. After the explosion other top officials either resigned or were reassigned. Center personnel at Huntsville, Houston, and Cape Canaveral squabbled over responsibility for the O-ring problem. NASA, the agency with the reputation for moving at top speed, seemed to be running out of steam. At times progress toward assessing and correcting the problems of the Shuttle seemed very slow.

With the Shuttle grounded it was obvious that the era of the expendable launch vehicle was by no means over. The air force, as it had feared, struggled to maintain the all-important reconnaissance-satellite system with a dwindling supply of aging Titan and Delta launch vehicles until a new generation of boosters, already on order, could be built and tested. Comsat/Intelsat and other commercial satellite firms gave their business to ESA and began to explore the possibility of using Soviet, Japanese, or Chinese boosters. The Reagan White House, anxious to encourage the growth of a U.S. commercial-launch industry, announced that in the future the Shuttle would fly only those payloads specifically designed for it or those that were vital to the nation's defense.

Three successive NASA administrators worked to get the organization back on track during the six years following the Challenger disaster. James C. Fletcher, who had served as administrator from 1970 to 1977, took over the post again in 1986. He was followed by veteran Shuttle astronaut Adm. Richard Truly, who turned over the reins to Daniel Goldin in 1991.

While the Americans were grounded, the Soviets continued to fly, extending their time in orbit. Mir, a third-generation space station, was carried into orbit on 20 February 1986 aboard a Proton SL-13. Larger, much better equipped, and more comfortable than its predecessors, the new station featured six docking ports. At a press conference on the morning of the launch, cosmonaut-spacecraft designer Konstantin Feoktistov explained that the Mir was "a base module for assembling a multipurpose permanently operating manned complex with specialized orbital modules for scientific and national economic purposes."[2] Cosmonauts would journey to Mir in a new Soyuz-TM (modified) spacecraft. The Soyuz-TM looked much like its predecessor, the Soyuz-T, but featured improved systems and could carry 332–442 pounds (150–200 kilograms) more equipment to and from orbit. The unmanned Progress robot spacecraft had also been updated.

Leonid Kizim and Vladimir Solovyov (Soyuz-T 15), launched on 13 March 1986, were the last crew to visit Salyut 7 and the first to board the orbiting Mir. It has been suggested that Soviet planners hoped to use the Salyut as an extension of Mir's core. In any event, the old station remained in orbit in 1992, waiting for another visit that never came.

The Soviets established a much heavier visiting schedule for Mir than had been the case with any of the Salyuts. By the spring of 1990 eleven crews had docked with the new station. The shortest crew stay was 151 days. By 1992 the record stood at 366 days in orbit for a single mission. Glavkosmos, the Soviet space agency created by Mikhail Gorbachev in 1985, had taken to capitalism with a vengeance. In December 1990 it sold a seat on a Mir-bound Soyuz to a Japanese television network for $12 million. A year later, the Austrian government paid $7 million to fly one of their nationals into space. By 1992 Glavkosmos had put Mir itself on the block—offering to sell the station to NASA.

The U.S. space agency was not interested in purchasing used Soviet hardware. On 29 September 1988 five U.S. astronauts flew back into orbit on mission STS-26. Over the next year and a half, the Shuttles would make ten more flights. The interval between missions, as long as three months in the spring of 1989, was gradually reduced to one launch every six to eight weeks. The false expectations of the early period of operations were gone, and the memory of the Challenger disaster began to fade.

Within four years news of the latest launch of a Space Shuttle was relegated to the back pages of the paper. The activities of human beings in space were of little concern to the producers of the nightly news. The same thing had occurred in the Soviet Union. The activities of a pair of bored and homesick cosmonauts who had spent a year in space paled into insignificance when compared to the drama of daily life in the Soviet Union of Gorbachev and Yeltsin.

Occasionally, usually when something went wrong, the old excitement would return. For two difficult days in 1992 the crew of STS-49 struggled to retrieve and repair the Intelsat 6 satellite while in orbit. For a few hours NASA Select, the nationwide cable-television channel operated by the space agency, drew a record number of viewers.

A few months later Western news reporters began to wonder whether the Russians, struggling to overcome administrative chaos following the breakup of the Soviet Union, would be able to organize the launch required to bring orbiting cosmonaut Sergei Krikalev home again. The ABC television news program "Nightline" featured a live interview with the cosmonaut, who was, in fact, never in any danger of being stranded.

If the collapse of the old order had created problems for the Russian space program, it had also opened the doors on an era of unprecedented cooperation. If the heirs of Korolev looked with envy on continuing support for what the Congress regarded as a reasonable U.S. space effort, NASA officials saw a way to use Russian equipment and experience in long-duration missions as a stepping stone to the creation of a next generation space station and a permanent human presence in orbit. The space race that had fueled the long-dreamed-of journey from Earth to the Moon had come to an end.

16 • At Home in Orbit

" **I** saw it first," Shannon Lucid recalls. "There were big thunderstorms out in the Atlantic, with a brilliant display of lightning. . . . The cities were strung out like Christmas lights along the coast—and there was the *Progress* like a bright morning star skimming along the top!!!" Lucid and her crewmates, Yuri Onufrienko, the commander of mission Mir 21, and flight engineer Yuri Usachev, watched the speck of light grow brighter until the deployed solar arrays of the approaching robot spacecraft became visible. "To me, it looked like some alien insect headed straight toward us." Lucid recalled. "All of a sudden I really did feel like I was in a 'cosmic outpost' anxiously awaiting supplies—and really hoping that my family did remember to send me some books and candy."[1]

Twenty-year-old Shannon Lucid earned her private pilot's license in 1963, ten full years before she graduated from the University of Oklahoma with a Ph.D. in biochemistry. Selected as an astronaut in 1978, she served as a mission specialist on four Shuttle flights between 1985 and 1993. In 1994 Robert "Hoot" Gibson, head of NASA's astronaut office, asked her if she would be interested in beginning Russian language instruction as the first step in preparing for a mission to the Russian space station, Mir ("peace"). She accepted at once. "For a scientist who loves flying," she noted, "what could be more exciting than working in a laboratory that hurtles around the earth at 17,000 miles (27,000 kilometers) per hour?"[2]

The spacecraft with which Shannon Lucid became so familiar was orbiting the earth at an altitude of 240 miles (400 kilometers), having grown to 120 tons with the addition of several specialized modules. The first element of the Mir station, the Core module, was launched from Baikonur on 20 February 1985. From the outset of the program it provided both living quarters and working space. "The floor of the operations area is covered with dark green carpet, the walls are light green and the ceiling is white with flourescent lamps," explains a NASA press release describing the Core module:

The arrangement of equipment and the interior finish of the working compartment are designed to reinforce this bottom-to-top orientation. The living area uses the same spatial orientation concepts, but soft pastel colors are used to imply a home-like atmosphere. The living area of the working compartment provides the necessities for long-term human missions. The living area contains a galley area with a table, cooking elements, and trash storage. Individual crew cabins, which include a porthole, hinged chairs and a sleeping bag are found next as one moves axially through the working compartment. The aft end of the working compartment contains the personal hygiene area with toilet, sink and shower.[3]

Mir was the first spacecraft designed to be expanded in orbit. Over the years a series of modules have been attached to the Core: Kristall (docking module, biological and materials testing), Kvant 1 (astrophysics), Kvant 2 (biology, earth observation, EVA port, life support), Spektr (laboratory space, power generation), and Priroda (remote sensing). Designed to permit a permanent human presence in space, Mir has hosted more than sixty cosmonauts and astronauts from Russia, the United States, France, Germany, Great Britain, Bulgaria, Japan, Canada, Kazakhstan, Austria, Syria, and Afghanistan. One of those cosmonauts, physician Valeri Polyakov, set the world's record for the longest stay in space: 438 days. More than eighty spacecraft have docked with Mir, from the Soyuz-TM and the U.S. Space Shuttles to the unmanned Progress capsules that deliver cargo to the orbiting station.

In the late 1980s the nations that composed the former Soviet Union still had high hopes of continuing the Russian tradition of leadership in space. They planned to continue flying the Soyuz-TM spacecraft into the next century. In addition, cosmonauts were scheduled to begin flying into orbit aboard the VKK ("air spaceship") in the 1990s.

The VKK, commonly known by the name Buran ("Siberian snowstorm"), would be carried aloft by the 3,934-ton (3,576-metric-ton) SL-17 Energiya

More than eighty spacecraft have docked with the Russian space station Mir. Here, the Space Shuttle Atlantis approaches in July 1995 to pick up the Mir-18 crew, which included astronaut Norm Thagard, the first American to live aboard the Russian station. (Courtesy of NASA)

booster, which was produced by Valentine Glushko and his associates at the old Korolev bureau. Introduced in 1988, the Energiya, 197 feet (60 meters) tall, is the first large Russian booster developed since the failure of the SL-15 and the final collapse of the Soviet lunar effort. It is also the first to use liquid-hydrogen-propulsion technology.

The Buran orbiter looked very much like a small-scale U.S. Space Shuttle. There were major differences, however, including the fact that the Energiya doubled as booster and the spacecraft main propulsion system. The small engines mounted on the spacecraft provided orbital maneuvering and attitude control capability. After a series of suborbital flights, Buran 1 made its only space flight in November 1988: two unmanned orbits followed by a safe, automated landing at Tyuratum. The program was designed to provide three to five launches a year, complete with a docking mission to a proposed Mir 2 space station.

That was the plan. Economic and political realities forced severe cutbacks in the space effort of the former Soviet Union. Some notion of the extent of this retrenchment is to be found in the fact that the 1992 budget for the Russian space program was 2.5 percent of the 1989 appropriation! "We're economizing in every way possible," explained veteran cosmonaut and head of the Russian manned space program Pytor Klimuk to an American reporter. "That's why all the lights are out in the hallways. Watch your step out there."

In spite of its promise as a next generation spacecraft, the development and operation of Buran proved to be far beyond the shrinking budget of the Russian space program. The project was officially canceled in 1993. Two years later, a pair of unfinished Buran orbiters were dismantled and the production facility was rededicated to the manufacture of diapers, syringes, and buses. Nor was an updated Mir 2 in the cards. Unwilling to break their record for the continuous habitation of space, the Russians sent one crew of cosmonauts after another to visit a rapidly aging Mir.

The possibility of a partnership with the United States and other nations to create an international space station, complete with financial support and generous contracts to the old design bureaus, offered real hope to beleaguered Russian space planners. Planning for the project began in January 1984, when President Ronald Reagan established the goal of developing a permanent human presence in orbit. By March 1989 NASA had signed agreements with Canada, the European Space Agency, and the government of Japan. By the spring of 1993, however, budget, management, and schedule difficulties had grown so serious that a complete reassessment of the project was undertaken in a series of meetings that became known as the Crystal City activity—for the Washington suburb where the study was centered.

NASA went into the exercise with the somewhat grandiose Reagan-era plan for Space Station Freedom, and emerged with International Space Station Alpha. While preserving the international partnerships, and as much of the planning for Space Station Freedom as possible, the new design would be implemented within the strictest budget constraints. NASA leadership struck a bargain with the Clinton administration. In exchange for a firm political commitment that would preclude the need for redesign or reductions, the agency agreed that the project could be accomplished with a flat budget of $2.1 billion per year, for a total of $17.4 billion.

With the program office established at the Johnson Space Center in Houston, the International Space Station (ISS) team was off and running. The station would be built by a consortium that had grown to sixteen nations. A total of forty-three U.S. and USSR launches over a period of six years would be required to carry the one hundred component modules and other major segments into orbit. An estimated one thousand hours of EVA time would be required for assembly. The finished product would have an interior volume equal to that of two jumbo jets.

A new phase of cooperation in space began in 1994, when the Soviet Union and the United States signed an agreement leading to joint operations aboard Mir. The United States agreed to pay some $400 million in "rent," and the Russians agreed to extend the life of the space station. Eleven Shuttle flights between 1994 and 1998 were identified as phase 1 of the ISS program. The goal was to use existing U.S. and USSR space hardware to build a body of joint operational and scientific experience in orbit. "We are laying a foundation for the construction of the International Space Station with these docking flights," NASA administrator Daniel Goldin commented. "Mir is proving to be an ideal test site for vital engineering research and expanding our knowledge of the effects of long-duration weightlessness on people."[4]

Clearly, however, this was as much an experiment in personal diplomacy and interpersonal relationships as it was an attempt to pursue science in space. "I think our primary mission is our relationship with the Russians," astronaut John Blaha once remarked. "So that working with my two crewmates is the primary mission. After that, of course, to conduct the experiments that we've been trained to conduct." Most of the individuals involved with phase 1 of the ISS program, whether Russian or American, would have probably have agreed.[5]

The program began in February 1994, when cosmonaut Sergei Krikalev flew as a NASA mission specialist aboard STS-60. One year later, with cosmonaut Vladimir Titov aboard, Discovery (STS-63) rendezvoused with Mir, closed to within forty feet, and checked docking and communication procedures. On 14 March 1995 astronaut Norm Thagard, an experienced astronaut who combined experience as both a marine fighter pilot and a physician, became the first American to fly aboard a Russian spacecraft when he launched from Baikonur as a member of the Mir 18 crew. He spent 115 days aboard the spacecraft and complained of a sense of isolation. "There are often snatches of the Russian news that are beamed up," he commented, "but that is my only access [to the world] unless I talk to some ham radio operators."[6]

Shannon Lucid was the second American to make her home aboard Mir. Thirty-five women had flown into space since Valentina Tereshkova had become the first woman to fly into space. Lucid would spend 188 days aboard Mir (22 March–26 August 1996), raising her total time in orbit (four Shuttle missions and Mir) to 222.1 days, and giving her more time in space than any other woman or any other American. She took up residence in the Spektr module, and her Russian colleagues continued to occupy cubicles in the Core. "Most mornings the wake-up alarm went off at eight o'clock [Moscow time]," Lucid recalls:

The first thing we usually did was put on our headsets and talk to mission control. . . . After the first com [communications] pass of the day, we ate breakfast. One of the most pleasant aspects of being part of the Mir crew was that we ate all of our meals together, floating around a table in the Base Block. Preflight, I had assumed that the repetitive nature of the menu would dampen my appetite, but to my surprise I was hungry at every meal.[7]

The days quickly fell into a routine of work, meals, and sleep, but Shannon Lucid did her fair share to prevent things from becoming monotonous. She celebrated each Sunday, for example, with pink socks and Jello. "The pink socks were found on STS-76," she explains, "and Kevin Chilton, the Commander, said they were obviously put on as a surprise for me, so I took them with me over to Mir and decided to wear them on Sundays." Lucid continues:

And the Jello? It is the greatest improvement in space flight since my first flight over ten years ago. When I found out there was a refrigerator on board Mir, I asked the food folks at JSC if they could put Jello in a drink bag. Once aboard Mir, we could just add hot water, put the bag in the refrigerator and, later, have a great treat. Well the food folks did just that and sent a variety of flavors with me to try out. We tried Jello first as a special treat for Easter. It was so great that we decided to share a bag of Jello every Sunday night. Every once in a while, Yuri will come up to me and say, "Isn't it Sunday?" and I will say "No, it's not. No Jello tonight!!!!"[8]

Remembering Norm Thagard's feeling of isolation, NASA officials provided Lucid with ample opportunities to send and receive daily reports from family

American astronaut Shannon W. Lucid catches up on her reading on Mir. Lucid spent 188 days aboard Mir, bringing her total days in orbit to 222, more time in space than any other woman or any other American. (Courtesy of NASA)

and friends, including unofficial amateur radio conversations when Mir was passing close to Houston. Every other week she was treated to a video conference with her family. Books, candy, and other treats provided by her family arrived with each Progress spacecraft. Lucid read fifty books while in orbit. Her scheduled return to Earth in August had to be postponed for six weeks while the Shuttle was grounded to enable engineers to study a problem with the solid propellant boosters. The delay called special attention to the fact that an American woman was setting records in space and guaranteed she would receive to a hero's welcome of a sort that had not greeted returning astronauts since Apollo 13.

Astronaut John Blaha replaced Shannon Lucid as the American aboard Mir (16 September 1996–22 January 1997). At that point Mir had been in operation five years longer than originally intended and was beginning to show its age. During a National Public Radio interview while the mission was underway, Blaha admitted that the older modules "look a bit used, as you could imagine something looking after people have lived in something in orbit for 10 to 11 years without having the advantage of bringing the vehicle home and letting it be cleaned up on the ground."[9]

Shannon Lucid would have agreed. Asked if she could offer any suggestions to the designers of the ISS on the basis of her experience aboard Mir, she
recommended that someone consider how to dispose of equipment and other
materials when they are no longer required in space. "On Mir, they have had
many expeditions over the years and they've brought various pieces of scientific equipment up," she noted. "There's no real good way of getting it back,
so they've got all this equipment up there. They've run out of places to
put it."[10]

As Jerry Linenger, John Blaha's replacement, was to discover, however,
Mir had far bigger problems than the need for a trash pickup. Linenger had
been on board about a month when, on 24 February 1997, a fire broke out
while the cosmonauts were attempting to change an air filter. Although the ultimate disaster was avoided, the Mir filled with smoke, forcing all three crew
members to don oxygen masks. It was a frightening moment, and others were
to follow. On 6 March a Progress supply ship was dumped after being unable
to dock, and the following day the failure of an oxygen generator left the crew
with a two-month air supply. A coolant leak on 4 April forced the crew to shut
down the primary system for scrubbing carbon dioxide from the cabin air,
and one week later Linenger admitted to reporters that the crew was suffering
from nasal congestion as a result of the coolant leak.

Those were only the big problems. In a letter of 20 March 1997 Linenger
explained to his young son, "Last night it got really, really, really dark in my
room—module Spektr. Lost all power." He continued:

I've been in dark places before, but this was un-earthly dark. Darker than any dark
I've ever seen. Dark is not even the proper word for it. And silent. So silent. Until then,
I hadn't really realized that you are constantly hearing some background ventilator/
machinery noise all the time. The silence was unfamiliar; even a bit surprising. Once
recognized, it was a very soothing, pleasant silence. Sounded nice. Of course, I couldn't
hang out in the quiet room. No ventilators means no air circulation. . . . Because of that
fact, you always pick a place to sleep where you can feel some air movement near your
head. If you don't, you'll end up within a self-generated carbon dioxide "bubble," wake
up panting with air hunger, and more likely than not, with a headache. By the way, that is
why I sleep upside down on the wall. . . . The ventilation is better near the floor—so I
want my head there.[11]

Linenger returned to Earth on 24 May 1975. His replacement, Mike Foale,
would have an even more difficult time aboard Mir. The worst accident in the
history of the program occurred on 27 June, when a seven-ton cargo module

smashed into the space station, destroying solar panels and punching a hole in the Spektr. "It was frightening for one or two seconds," Mike Foale recalls. "The first thought was are we going to die instantly because of the air rushing out so fast we can't control it?"[12]

The problems were far from over. On 27 June the Mir computer "disconnected" from the control system when batteries ran low. The gyroscopes that orient the spacecraft shut down on 3 July, and two days later cosmonauts report an "unidentified substance" leaking from the damaged Spektr. On 17 July a crew member accidentally disconnected a vital cable, causing the Mir to lose power. Two oxygen generators broke down on 5 August, forcing the crew to resort to special oxygen canisters, and twenty days later both primary and backup oxygen systems failed (generators were restored within a few hours). On 18 September the Mir computer failed for the third time since July, and the crew was forced to turn off all equipment not directly related to life support.

Americans, watching the list of problems with Mir growing ever longer

Astronaut Kevin P. Chilton *(center)* **greets Mir cosmonauts Yuri I. Onufrienko** *(left)* **and Yuri V. Usachev on the flight deck of the Space Shuttle Atlantis, March 1996. Over the next few years the collapse of the Russian economy made funding such ventures increasingly difficult. The last American aboard Mir, Andy Thomas, departed on 12 June 1998. (Courtesy of NASA)**

and more serious, became increasingly concerned. As early as April Florida representative Dan Weldon, initially a supporter of the project, expressed his opposition to sending another astronaut to Mir simply to "spend months being an assistant Mr. Fixit." On 11 July 1998 Wisconsin congressman F. James Sensenbrenner, chairman of the House Committee on Science, asked NASA's inspector general to assess U.S. participation in the Mir effort and answer some specific questions relating to the safety, productivity, and cost effectiveness of the effort. In his reply, the IG admitted to serious problems in all areas of concern.

The problems of Mir were front-page news in Russia as well. In a press conference that surely would never have been held in the Soviet era, returning cosmonauts Vasily Tsibliyev and Alexander Lazutkin complained that the media and the government had used them as scapegoats for the string of catastrophes that had struck Mir. "The cause lies in the problems back on Earth," Tsibliyev commented. "It's connected with the economy and the state of things in general. It's impossible to procure many things which are vital to the station, due to the fact that they are either not manufactured any longer . . . [or] overpriced."[13]

As a CNN news release explained, the House Science Committee "opposed sending any more Americans into orbit, but left the final decision—and the responsibility if anything went wrong—to NASA Administrator Dan Goldin."[14] Weighing all of the factors, including the opinions of NASA technical staff, Goldin decided to continue the cooperative effort. "Despite my objections," Congressman Weldon explained, "the administrator chose to go ahead and continue the program. There haven't been any more major mishaps , and I am grateful, thankful to the Lord for that."[15]

In spite of what seemed to be overwhelming difficulties, the project that NASA identified as phase 1 of the ISS program moved forward to completion. David Wolf replaced Mike Foal (25 September 1997–31 January 1998) and was followed in turn by Andy Thomas (22 January 1998–12 June 1998), the last American to live and work aboard Mir. Anxious to move forward with the space station, NASA officials put matters in the most positive context as they closed the books on phase 1. "It's very difficult to imagine beginning Phase 2 without doing what we've done during the Shuttle-Mir program," one agency leader commented. "Getting to know how to operate in space, getting to know how to work with an international partner, getting to understand the Russian way of doing business, has been critical to beginning the [station] operations in space."[16]

Having said that, NASA leaders were also quick to agree that enough was enough. Concerned that the Russians would not be able to continue operating Mir and meet their considerable obligations to the International Space Station effort, NASA and its other international partners began forcefully arguing that the Russians should phase out the Mir program. "It's a resource problem," noted Keith Reiley, a space station manager. "We're concerned [that] if there is a problem, ISS will get the short end of the stick."[17]

Somewhat reluctantly, the Russians agreed. They could not produce enough Soyuz vehicles to rotate Mir crews and ferry personnel to the ISS. Plans called for Mir to continue hosting crews until the early summer of 1999. Progress supply ships, the engines of which had been used to prevent Mir from sinking into a lower orbit, would be used to initiate a slow descent.

A month before scheduled reentry in July/August 1999, the station will have dropped into an elliptical orbit that will carry it to within one hundred miles of Earth at the low point. At a precisely timed moment, a Progress burn will send the spacecraft streaking back to Earth in a remote area of the Pacific, where the larger chunks of debris that survive a fiery reentry will not do any damage. The event will mark the end of an era, when the Russian bear made its mark in space, from the glory days of Sputnik and Yuri Gagarin to the pioneering effort to achieve long-duration stays in orbit.

17 • Robot Servants and Explorers

The National Aeronautics and Space Administration celebrated its twenty-fifth anniversary in 1983. The agency commemorated the occasion with museum exhibitions, films, and special publications. In all of the excitement an enormously significant milestone passed almost unnoticed: At precisely 5:00 A.M. Pacific time, on 13 June 1983, Pioneer 10 officially became the first product of human civilization to leave the solar system.

Launched eleven years earlier, on 3 March 1972, Pioneer already had piled up a spectacular record. It was the first spacecraft to travel beyond the orbit of Mars, crossing the asteroid belt to return the first closeup images of Jupiter. Its primary mission complete, Pioneer 10 flew across the orbits of the outer planets and into the infinity of interstellar space. NASA equipped humanity's first traveler to the stars with a suitable calling card: a small metal plaque showing a man and a woman, a diagram of the solar system, and a map locating the Sun in relation to several galactic pulsars.

Although public attention usually focused on the dramatic activities of the astronauts and cosmonauts, the men and women who design, build, and fly satellites and planetary probes were laying the foundations for the *real* spaceflight revolution. They seldom make headlines, yet their work has enabled scientists to probe the deepest mysteries of the universe and has touched the everyday lives of people around the world in the most profound ways.

The world's first spacecraft were scientific satellites designed to impress the world and explore the complex environment of particles and fields in our immediate cosmic neighborhood. Between 1958 and 1964 Pioneer and Mariner probes, along with a series of Explorer scientific satellites, as well as various Soviet unmanned spacecraft, measured and mapped the flow of high-energy particles streaming out from the Sun to be captured in Earth's magnetic field. Scientists wondered about the relationship between an increase in solar flares and sunspots and the intensity of the aurora borealis, as well as about disturbances in radio communications being bounced off ionized layers in the upper atmosphere.

The first satellites and space probes unlocked the mystery. They confirmed the existence of what became known as the solar wind: a flow of particles streaming out from the Sun in all directions. The various spacecraft proved that the energy moves at various speeds, depending on the level of solar activity, and contains electrically charged particles. When the solar wind encounters a magnetic field surrounding a planet like Earth, it is deflected, flowing around the magnetic field like water around a partially submerged rock. The complex area of interaction between the solar wind and Earth's magnetic field, from the "bow shock wave" closest to the Sun to the tip of the tail extending beyond the orbit of the Moon, is known as the magnetosphere.

But not all of the solar wind is deflected around Earth. Some particles are trapped and held in two doughnut-shaped belts by Earth's magnetic field. Both belts are named after their discoverer, James Van Allen, of the University of Iowa. The inner radiation belt, composed mostly of protons, is centered some 2,500 miles (4,025 kilometers) above the equator. The outer electron belt is located 10,000 miles (16,100 kilometers) above the surface. The study of the solar wind, magnetosphere, radiation belts, and other aspects of the relationship between high-energy particles and magnetic fields in space continues today.

With a record of early successes behind them, NASA space planners began launching larger, more complex scientific satellites in the early 1960s. Known as observatory spacecraft, they weighed five to ten times as much as the original Explorers and could be controlled in orbit and pointed at targets to a degree impossible with their predecessors. Between 1962 and 1975 the agency successfully launched eight orbiting solar observatories, six orbiting geophysical observatories, and three orbiting astronomical observatories. Each observatory series consisted of a basic spacecraft design that could carry up to twenty experiments per flight. In addition NASA flew a series of three specialized biosatellites, designed to study the effects of space conditions on living organisms, and Pegasus, which gauged the amount of micrometeoroid activity in near-Earth space.

The new generation of scientific spacecraft produced a wealth of information, and provided astronomers, geophysicists, and astrophysicists with a revolutionary new set of tools. Orbiting Astronomical Observatory 2, launched on 7 December 1968, took the first ultraviolet photographs of stars and provided hard evidence for the existence of the incredibly dense collapsed stars known as black holes. The Uhuru satellite of 1970 searched the sky for X-ray sources and found 127 new ones. It also identified three new pulsar stars. Orbiting Solar Observatory 7, launched in September 1971, caught the beginning of a solar flare on film for the first time, and it discovered polar caps on the Sun. LAGEOS (laser geodynamics satellite), launched on 4 May 1976, investigated the most down-to-earth problems imaginable: the earth's crustal movements and continental drift. Astrophysics, however, remains the single most important area of satellite research.

Dozens of satellites continue the study of elementary particles and electromagnetic fields in near-Earth and interplanetary space. Explorer spacecraft of the interplanetary monitoring platform (IMP) series kept track of changes in the near-Earth radiation environment during an eleven-year solar cycle. The Sun continues to draw a great deal of attention from such satellites as the Solar Maximum Mission (1980–89), which studied the corona and solar flares.

Still other satellites look out into the universe. During the past two decades, orbiting scientific platforms have mapped the heavens across the spectral band from the infrared to the ultraviolet. These spacecraft have had a profound impact on our understanding of the nature and origins of the cosmos. One of these satellites, the Cosmic Background Explorer (COBE), launched on 18 November 1989, provided infrared and microwave sky maps that helped to confirm the big bang theory of the origin of the universe. It was, any number of authorities noted, one of the most fundamentally important space-astronomy missions ever undertaken.

During the late 1970s NASA was planning to cap its program of orbital space science with a new generation of observatory spacecraft. In 1983 the first of these, originally known as the Large Space Telescope (LST), was renamed the Hubble Space Telescope in honor of eminent American astronomer Edwin Hubble. The spacecraft would observe objects twenty-five times fainter and ten times more distant than were visible with the largest telescopes on Earth. The great mirror of the Hubble telescope would gather in the light generated fourteen billion years earlier, on the very edge of the known universe. The forty-three-foot (thirteen-meter) observatory, which drew on experience with similar optical instruments flown in reconnaissance-satellite programs, suffered long delays while problems discovered during ground testing were corrected. Originally scheduled for a 1986 launch, the project was grounded indefinitely when Shuttle flights came to a halt following the Challenger disaster.

The Hubble Space Telescope finally was flown from California to Cape Canaveral aboard a U.S. Air Force Lockheed C-5A in October 1989 and launched into orbit on 24 April 1990 with the crew of STS-31. NASA's prestige, which still had not recovered from the loss of the Challenger crew four years earlier, suffered a staggering blow as ground controllers and scientists began to analyze the data being returned by their highly touted spacecraft. They discovered a two-micron grinding error around the edge of the primary mirror. Most of the planned observations in the ultraviolet portion of the spectrum would not be affected, but the performance of the telescope in the visible-light range, critical to some of the most important studies planned for the instrument, would be little or no better than that of Earth-bound telescopes.

An independent investigating board determined that the instrument used to check the work had been improperly calibrated. Other tests had revealed the problem, but they were disregarded. NASA had chosen not to perform a complete test of the optical system, in part as a cost-saving measure. Public reaction ranged from incredulity to ridicule. Seldom in its thirty-two-year history had NASA's stock sunk so low. But America's most visible agency had grown accustomed to recovering from apparent disasters. In December 1993 the crew of STS-615 paid a long-awaited service call to the orbiting Hubble telescope. Millions of television viewers around the globe watched as the astronauts struggled to capture the huge satellite, and they remained on the edge of their seats during the long, hard hours of EVA as space-suited astronauts performed difficult repairs and replaced damaged or malfunctioning parts.

The mission could not have been more successful. The proponents of manned space flight could point to a perfect case in which astronauts had saved a critically important satellite with the whole world watching. More-

The partially deployed Hubble Space Telescope moves out from the Space Shuttle Discovery on 24 April 1990. A two-micron grinding error, which drastically reduced the quality of the Hubble's images, was repaired during a dramatic space service call in October 1993. (Courtesy of NASA)

over, the Hubble Space Telescope proved to be even more successful than its adherents had predicted. The instruments returned a wealth of data and a treasure trove of stunning images: from a moment-by-moment visual record of the pieces of Comet Shoemaker-Cannon during their fiery entry into the Jovian atmosphere to incredible, glowing images of clouds of dust and gas at the edge of the galaxy.

The second of the large observatory satellites, the Gamma Ray Observatory (GRO), was carried into orbit by the Space Shuttle on 5 April 1991. The Chandra Advanced X-Ray Astrophysical Facility, named in honor of Subrahmanyan Chandrasek, cowinner of the 1983 Nobel Prize for his work on the death of stars, was launched in July 1998 by the STS-93 crew, commanded by Lt. Col. Eileen Collins, the first woman to command a Shuttle mission. Plans for the Space Infrared Telescope Facility (SIRTF) are still under discussion. In four short decades scientific satellites have extended our vision very close to the edge of the universe.

The applications satellite programs pioneered by NASA represent one of the great success stories of the space age. From the outset the agency sought to limit its role to the identification, development, and demonstration of potentially useful space-based technologies. With that accomplished, NASA would turn over the day-to-day problems of maintenance and operation of a mature system to an agency or a corporation better equipped to meet the needs of users. That was the plan. In practice things were never that simple. Take the case of the communications satellite, or comsat, for example.

As early as 1945 English author and space-flight authority Arthur C. Clarke had suggested that a series of microwave-radio relays in Earth orbit could serve as the basis for a global-communications system. The first step toward the achievement of that goal was taken in December 1958 with the launch of the Advanced Research Project Agency's SCORE (signal communication by orbiting relay equipment) satellite, which broadcast, among other things, a prerecorded Christmas greeting from President Dwight Eisenhower. SCORE was followed by Echo 1, NASA's first comsat, launched on 12 August 1960. A one-hundred-foot (thirty-meter) inflatable balloon, Echo was a passive target, radio signals aimed at it reflecting back to Earth. Courier 1B, the first active comsat capable of receiving, amplifying, and rebroadcasting messages from Earth, was launched by the U.S. Army on 4 October 1960.

Major communications companies were quick to recognize a potentially profitable revolution in the making and anxious to establish themselves in the field. The first active civilian comsat, Telestar 1, was developed by the American Telephone and Telegraph Company (AT&T) and launched by NASA on 10 July 1962. Relay 1, the nation's third active comsat, produced by the Radio Corporation of America (RCA), followed on 13 December 1962.

The Relay and Telestar series were technology demonstrators, transmitting the first civilian-television broadcasts from America to Europe and Japan, but they flew in relatively low orbits and had to be constantly tracked. The Hughes Aircraft Corporation pioneered operations at synchronous altitude with the development of Syncom 2 and 3, launched by NASA on 26 July

1963 and 19 August 1964, respectively. Orbiting 22,230 miles (35,790 kilometers) above the equator, the Syncoms traveled at a speed precisely matching that of the rotation of the earth, so that they remained constantly positioned over one spot, enabling each to cover one-third of the earth's surface.

With the success of the Syncoms, the world stood at the edge of the most significant revolution in communications technology since the invention of radio. There was, however, a problem. Although the U.S. government did not want to become involved in the operation of an international space-communications system, the prospect of a single private corporation using the fruits of NASA research to establish a monopoly over a revolutionary new technology was not acceptable. Congress, along with officials of the Federal Communications Commission and the Department of Justice, forged a solution to the dilemma. On 31 August 1962 after extended congressional debate, President John Kennedy signed a bill creating the Communications Satellite Corporation (Comsat), a unique organization that would be half owned and controlled by the government and half owned and controlled by large communications companies willing to invest in the venture.

Comsat flourished. International agreements led to the creation of the International Telecommunications Satellite Consortium (Intelsat) in August 1964, with Comsat serving as the American member. NASA launched the organization's first satellite, Early Bird, on 6 April 1965. By 1993 membership in Intelsat had grown to 125 nations, and the consortium was operating a successful and enormously profitable worldwide communications-satellite network. Two-thirds of all international telephone calls and virtually all transoceanic television service is broadcast via Intelsat channels. International financial organizations, news services, and the United Nations also are heavy users of the network.

Comsat remains the U.S. signatory to Intelsat, controlling roughly 22 percent of the international organization's shares. It has attracted investors from a wide range of U.S. communications companies. The company provides essential technical services for its international partners, working with established aerospace companies to design and build new generations of Intelsat spacecraft, purchasing launch services from NASA (and eventually from other nations), and selling time and channels on its satellites to broadcast companies and other users.

The Comsat International Division manages fifteen communications companies that are of central importance to the economies of thirteen nations around the globe. The Comsat laboratories pursue research and development projects that keep the company at the leading edge of communications technology. As of 1998 Comsat Mobile Communications maintains a distinct global system of nine Inmarsats that provides maritime, aeronautical and land mobile, and personal communications services.

Since the creation of Comsat and Intelsat, NASA has limited its involvement in communications-satellite technology to areas of research in which the profit margin is low and the potential for public service high. Three of the six applications technology satellites (ATS) launched between December 1966 and May 1974 explored new facets of communications satellite technology. ATS-6 was particularly noteworthy. Capable of broadcasting to small and inexpensive ground stations, it provided experimental public-health programming to remote areas in the United States and India.

NASA's second great applications success, the meteorological satellite, presented the agency with a very different set of organizational problems. NASA orbited ten TIROS (television infrared observation satellites) between 1960 and 1965, which returned more than five hundred thousand images of the earth's cloud cover. Prior to the launch of TIROS 1, the U.S. Weather Bureau agreed to analyze the cloud-cover pictures and other information returned by the satellite. Bureau officials were stunned by the quality of the photos and information beamed back from space. Within a few weeks, cloud-cover maps and forecasts based on the TIROS data were being distributed nationwide.

Having tasted success, NASA engineers were eager to push ahead with improved imaging and remote-sensing technologies that would be flown aboard a new generation of Nimbus satellites. The Weather Bureau, completely satisfied with the existing TIROS data and unwilling to invest in advanced technology that it did not need, threatened to withdraw from its agreement with NASA in favor of cooperation with other TIROS users in the Department of Defense. The DoD feared a critical gap in weather coverage between the TIROS and Nimbus flights. Anxious to recapture some of the ground it had lost from key space programs that had been transferred to NASA, the military services now were contemplating their own system of weather satellites. Faced with the prospect of losing its only customers for a highly touted service, NASA agreed to identify Nimbus as an experimental system and to create a TIROS operational system (TOS) to meet the needs of users such as the Weather Bureau. At last there was a meeting of the minds.

Astronauts wrestle the wayward Intelsat 6 communications satellite into the cargo bay of the Space Shuttle Endeavor, May 1992. The rescue mission, involving four space walks totaling more than twenty-four hours, tested methods used in the capture and repair of the Hubble Space Telescope the following year. (Courtesy of NASA)

NASA would continue to expand the edges of the technology and to provide launch services; the user agencies would shape the operating systems to suit their own needs.

The final two satellites in the TIROS series, TIROS 9 and TIROS 10, conducted experiments and tested equipment scheduled to be flown on the new-generation TIROS operational system (TOS-A to TOS-G). Once in orbit, the TOS vehicles would be called ESSA 1, ESSA 2, and so on, for the Environmental Science Services Administration, the federal agency that funded their development and was slated to operate them in space. On 11 December 1970

ITOS 1 (improved TIROS operational system), the first in yet another upgrade of this meteorological spacecraft, was launched. The TOS-A to TOS-H satellites were operated by a new federal agency, the National Oceanic and Atmospheric Administration (NOAA) and became known as NOAA satellites following launch.

The agreement between NASA and its user agencies held. ESSA and NOAA managed the operational weather-satellite program, and NASA continued to build and launch experimental meteorological satellites. Seven Nimbus spacecraft launched between 1964 and 1978 returned the first images of Earth at night and tested a variety of instruments that opened the way to twenty-four-hour satellite weather coverage. The synchronous meteorological satellites (SMS) launched in 1974 and 1975 were transferred from NASA to NOAA and led to a series of eight GOES (geostationary operational environmental satellites) vehicles that have been built for and launched by NASA, with funding from NOAA, between 1975 and 1996.

With the meteorological satellite program underway, NASA scientists and engineers recognized that the ability to take high-resolution images of the earth's surface in a variety of spectral bands would prove useful in solving a range of problems. The value of weather and communication satellites was readily apparent. In the case of orbital remote-sensing, however, NASA had to identify potential users and convince them of this service's value. Two potential customers for an operational Earth-observation system, the Departments of Agriculture and the Interior, requested instruments producing different sets of data, whereas state and local governments, NASA's target audience, showed little initial interest in the project. The resulting vehicles represented a compromise that did not suit the precise needs of any of the users but did offer a convincing demonstration of the broad utility of satellite imagery and data.

Landsats 1, 2, and 3, launched between July 1972 and March 1978, returned millions of images of great value to geologists, foresters, cartographers, agricultural planners, land- and water-resource managers, and city planners. Landsat data was used to search for new oil and mineral deposits, to map the spread of pollution, to predict crop production, to calculate the amount of available water based on snowmelt, and to assess the damage caused by earthquakes, forest fires, and floods.

Landsat data, telemetered back to Earth, led to stunning discoveries, ranging from an unsuspected geological fault in Great Britain to sources of pollution in the Chesapeake Bay. The Brazilian government realized that remote towns actually were located tens of kilometers from the position shown on the most up-to-date maps. Maps of the Caspian Sea had to be redrawn, and a South American lagoon was found to be five times as long as previously supposed.

NASA was hard pressed to deal with the huge quantity of information pouring back to Earth and was unable to encourage potential users to explore the benefits of space imagery. As had been done with weather and communications satellites, NASA officials had planned to demonstrate Landsat technology and then turn the program over to another agency or private corporation capable of marketing the data and funding the creation of an operational system.

In 1979 NOAA agreed to manage the Landsat system and assist users. The organization accepted operational control of Landsat 4, orbited in July 1982, and funded the development of Landsat 5, which was launched on 1 March 1984. NOAA officials, however, were anxious to find a commercial operator for the Landsat system. In September 1985, the agency signed a contract with EOSAT, a new firm based on a partnership between Hughes Aircraft (later General Motors) and the RCA Corporation (later GE). With funding from the Department of Commerce, EOSAT would handle the development of next-generation sensors and spacecraft and would market existing Landsat data.

The effort to commercialize the Landsat program, however, has been less than successful. In 1989 a reduction in congressional funding threatened to bring the program to a halt. Few knowledgeable specialists doubt the value of Landsat, but the program's potential to be economically self-sustaining remains undeveloped. Over the past two decades, specialized Earth-observation spacecraft, including Seasat, which initiated specialized ocean observation in 1976, have had an enormous impact on environmental research.

Some of the most important applications satellite programs of the past forty years remain secret today. The desire to sample the military advantages of an orbital platform was one of the most compelling reasons for pursuing space flight in the first place. From 1957 to the collapse of the Soviet Empire and the end of the Cold War, the military services of the United States and the Soviet Union built and flew their own series of increasingly sophisticated reconnaissance satellites, as well as military spacecraft for communications, navigation, early warning of a ballistic-missile attack, the detection of nuclear explosions, and a number of additional tasks. The details of many of these programs remain classified. Teasing out information on the reconnaissance-satellite effort, the very existence of which the U.S. government would not admit until 1995, kept investigative reporters covering the aerospace industry busy for thirty years. Whatever the technical details of these satellites, it is clear that they have revolutionized the business of intelligence gathering. By dramatically reducing the possibility of a surprise attack by land, sea, or air, these all-seeing eyes in the sky made a major contribution to keeping the peace in the nuclear age.

One class of military spacecraft, the navigation satellite, or navsat, also has played a major role in world commerce. Transit 1B (13 April 1960), the first navsat, was designed to allow Polaris submarines to fix their positions to within one-tenth of a mile so they could navigate and fire their nuclear missiles accurately. The success of the program led to the development of in-

Similar to Echo 1, this 135-foot passive communications satellite provided an orbital target for reflecting radio signals back to Earth. Folded for launch the satellite fit into the 41-inch canister shown in the foreground. (Courtesy of NASA)

creasingly sophisticated systems that became the basis for virtually all military position finding. The global positioning system (GPS), which began operation in 1994, is the latest navigation system for a wide variety of military and commercial users.

The complex part of the system is aboard the series of GPS satellites circling the globe; a lightweight, hand-held unit provides absolutely accurate positioning. By the spring of 1998 more than one million GPS receivers were being sold annually. They are carried by ships, from the largest supertankers to small sailboats, and by all varieties of aircraft.

From 1957 to 1990 the Soviet Union designed, built, and flew its own series of scientific and applications satellites. In fact, the Russians succeeded in placing 2,256 payloads in orbit, more than twice the number of orbital missions conducted by NASA and the U.S. military space programs (935). With the Shuttle grounded in 1986 and 1987, the Soviet orbital-launch rate was fifteen times higher than that of the United States. As late as 1990 the USSR orbited seventy-five payloads as opposed to twenty-seven for the United States. This striking difference in the frequency of launches stems in part from the Soviet preference for critical applications satellite systems requiring more spacecraft in lower orbit and the need for a more frequent replacement of satellites in the system.

On 26 November 1965, with the launch of its A-1 (Asterix) satellite, France became the third spacefaring nation. By 1990 seven other nations or consortiums had succeeded in orbiting at least one payload: Japan (forty-one), Europe (thirty-five), China (twenty-eight), India (three), Israel (two), Australia (one), United Kingdom (one). The leading international aerospace journal *Interval* reports that a total of forty-six nations were operating domestic space programs in 1991. These ranged from nations such as Canada that design and build satellites for domestic use but purchase launch services from another space program to countries that regularly produce instruments, components, or complete systems for other space programs.

It is not often easy to predict the social changes that will flow from a new set of technologies. The rocket pioneers who dreamed of realizing human destiny beyond in the stars would surely have been surprised to discover that although the first-generation citizens of the space age have found it hard to sustain a high level of interest in the activities of astronauts and cosmonauts, Earth satellites have revolutionized the way in which human beings communicate, predict the weather, find their way, understand their environment, and probe the secrets of the universe. "Thanks to a few tons of electronic gear twenty thousand miles above the equator," an optimistic Arthur C. Clarke remarked as early as 1965, "ours will be the last century of the savage; and for all mankind, the Stone Age will be over."[1]

Some of the most exciting discoveries of the space age have been made by robot planetary explorers. Early attention focused on Earth's closest neighbors, Venus and Mars. On 14 December 1962 the Mariner 2 spacecraft flew past Venus at a distance of 21,625 miles (34,816 kilometers). Although not equipped with cameras, the vehicle demonstrated that, unlike Earth, Venus has a very weak magnetic field and no radiation belts. In addition, the first successful U.S. interplanetary explorer determined that the surface temperature of the planet was a very high 800 degrees F (427 degrees C).

The next U.S. visitor to Venus, Mariner 5, flew past the planet on 19 October 1967, one day after the Soviet's Venera 4 probe achieved the first landing on the surface. Between them, the two spacecraft indicated that Venus has an atmosphere composed roughly of 87 percent carbon dioxide, an even higher surface temperature than indicated by Mariner 2 (980 degrees F, or 527 degrees C), and a surface atmospheric pressure perhaps one hundred times greater than that of Earth.

A third U.S. spacecraft, Mariner 10, flew past Venus in February 1974, returning detailed images of the dense clouds shrouding the surface of the inhospitable planet. Using a gravity assist acquired as it swung past Venus, Mariner 10 continued on to Mercury. It completed three passes over the surface of this desolate, innermost planet in March and September 1974, and in March 1975 provided the first closeup of the face of Mercury. NASA's next attempt to probe the mysteries of Venus came with the launch of Pioneer Venus 1 and 2 in 1978. The first spacecraft went into orbit on 4 December and would continue to return data and images of Venus for the next decade. Pioneer Venus 2 arrived at its destination five days later, on 9 December 1978.

The second spacecraft consisted of a large transporter structure, or bus; one large probe; and three identical small probes. Plunging toward the planet, the spacecraft returned information on the upper atmosphere, discharged its probes, and was destroyed by the heat of entry. The four probes transmitted data during their descent to the surface. Although not designed to survive impact, one of the probes continued transmitting for sixty-seven minutes after it crashed. The Soviets continued their exploration of Venus with the Venera 9 to Venera 16 missions from 1975 to 1986. Venera 6 returned the first panoramic photographs taken on the surface of the planet. Subsequent missions returned higher quality photos. The later model Venera craft used side-looking radar and thermal sensors to map the surface.

Two Vega (a word formed from the contraction of Venera-Galilei, or Venus-Halley) spacecraft were launched by Proton boosters on 15 and 21 December 1984. Designed for a dual mission to Venus and Halley's comet, the twin probes were the product of a genuinely international program involving the USSR, Austria, France, Bulgaria, Czechoslovakia, Hungary, East Germany, West Germany, and Poland.

Sweeping past Venus four days apart in June 1985, each Vega released a capsule containing a lander and an instrument package that would be held aloft in the atmosphere by the first balloon ever deployed on another planet. The two landers failed to operate as long as planned, but they did return information on the composition of the soil. Tossed about by storms in the Venusian atmosphere, the balloon-borne instruments covered a distance of more than 6,200 miles (10,000 kilometers) over the surface of the planet. During the second half of their mission, the Vega spacecraft obtained spectacular images of Halley's comet.

The latest visitor to Venus, NASA's Magellan spacecraft, contained a fair number of spare parts left over from Voyager and Galileo planetary probes. Launched from the Shuttle Atlantis on 4 May 1989, Magellan was equipped with radar instruments enabling it to map the planet's surface features through the clouds. For the first time, the clouds that shrouded the surface of the "evening star" from view were stripped away. Venus, the great mystery of the solar system since the invention of the telescope, was transformed into one of the best known and most thoroughly mapped objects in the solar system.

Of all the planets, Mars seems most intriguing. It is not surprising that NASA chose to concentrate on Earth's mysterious red neighbor. In July 1965 Mariner 4 became the first spacecraft to conduct a flyby of Mars, and Mariners 6 and 7 followed in July and August of 1969. The narrow strips of surface photographs transmitted back to Earth by the three spacecraft were something of a disappointment to those who had hoped for signs of a vanished civilization. The planet looked bleak and as pockmarked with craters as the surface of the Moon. As with Venus, Mars lacked a magnetic field and radiation belts, but it did support a thin atmosphere.

While the Soviets concentrated on Venus, NASA returned to Mars with

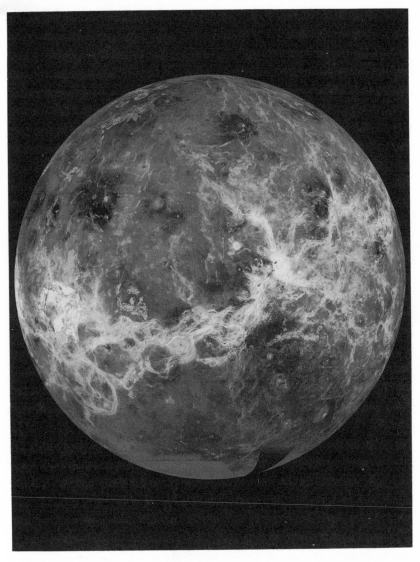

Magellan's radar instruments, capable of penetrating the planet's dense sulfurous clouds, sent back this first portrait of the surface of Venus. The circular features, thought to be unique to Venus, are ridges and troughs believed to form over hot plumes rising from the planet's interior. (Courtesy of NASA)

Mariner 9 in November 1971. As luck would have it, the first spacecraft actually to orbit another planet arrived in the middle of a gigantic dust storm. When the surface of Mars began to emerge two months later, scientists were delighted to find a very different world from the one suggested by the flyby spacecraft. One of the first features to appear was an enormous volcanic crater, Olympus Mons, 18 miles (29 kilometers) high and more than 300 miles (483 kilometers) in diameter.

Despite looking dead and barren, Mars once had been the scene of volcanic activity on a scale so enormous that this one crater dwarfed anything to be found on Earth. The planet's surface was broken with fracture patterns indicating massive geological activity. There were huge canyon systems stretching across an area equal to the width of the United States, and what appeared to be dried-up river beds. No signs of life were visible from orbit, but lost civilization or no, Mars had become an interesting place once again.

The next logical step was Viking. The Viking program of 1976 was not only the high point of NASA's Mars exploration effort but also in many ways the most ambitious of all interplanetary missions attempted up to that time. It consisted of two pairs of spacecraft, each pair consisting of one Viking orbiter and one lander. The landers would ride piggyback on the orbiters during the journey to Mars. Launched a few weeks apart, on 20 August and 9 September 1975, they flew to Mars and went into orbit, surveying preselected landing spots to be sure they were suitable.

The Viking 1 lander touched down in a basin known as the Chryse Planitia on 20 July 1976. On 3 September the second lander descended to the surface of the Utopia Planitia. For the next year the two landers photographed every aspect of the surface within their view, kept tabs on local weather conditions, analyzed the composition of the atmosphere, and dropped scoopfuls of soil into an on-board chemical laboratory that reported its constituent elements and searched for traces of organic material. At the same time, the orbiters were mapping the planet and its moons and relaying information and instructions between scientists on Earth and the two landers. The Viking project did not discover life on Mars, but it did demonstrate the extent to which sophisticated robot spacecraft could thoroughly explore another world.

On 4 July 1997 the small and comparatively inexpensive Mars Pathfinder came streaking down through the thin Martian atmosphere and bounced onto the surface of the Red Planet, cushioned by a set of inflatable, shock absorbing balls. Once the craft had come to rest, the first order of business was to

The Viking 1 camera reveals sand dunes and large rocks on the surface of Mars on 23 July 1976. The late afternoon Sun is high in the sky over the left side of the photo. (Courtesy of NASA)

transmit status reports of on-board systems, information on the local weather, and photos of the neighborhood.

With the preliminaries out of the way, a twenty-five-pound, six-wheeled robot explorer named Sojourner rolled off the spacecraft and went to work. Named by a twelve-year-old Bridgeport, Connecticut, girl in honor of the African American abolitionist heroine Sojourner Truth, the little rover and the stationary Pathfinder remained active until early November, when the depletion of the batteries led to a loss of contract. During an active life of four months, the spacecraft returned 2.6 billion bits of information, including 16,000 photos from the lander and 550 images from the rover, fifteen chemical analyses of rocks in the vicinity, and an entire archive of information on local environmental conditions. "Done quickly and within a very limited budget," David Baltimore, the president of Caltech explained, "Pathfinder sets a standard for 21st century exploration."[2]

Given its proximity and potential for interesting discoveries, Mars will continue to hold the attention of mission planners. The Mars Airborne Geophysical Explorer (MAGE) promises to be one of the best publicized planetary missions proposed for the first decade of the twenty-first century. The high point of the mission is scheduled for 17 December 2003, the one-hundredth anniversary of the Wright brothers' historic flights at Kitty Hawk, when a thirty-pound winged aircraft developed by the Naval Research Laboratory will be deployed from an entry vehicle descending through the Martian atmosphere. Spreading its wings to a span of thirty-two feet, the automated MAGE, named Kitty Hawk, will orient itself and fly for three hours along the course of the great Martian rift valley, the Valles Marineris. Cruising at an altitude of thirty thousand feet, Kitty Hawk will be armed with an array of geophysical instruments and both still and video cameras. Close to the end of the flight, the spacecraft that released Kitty Hawk will pass overhead, collecting information from the aerial explorer and relaying it back to Earth. If all goes as planned, it will be the most suitable commemoration of a century of flight imaginable. Wilbur and Orville would approve.

The outer planets were the most distant targets for NASA's automated explorers. Pioneers 10 and 11, designed to conduct flyby surveys of Jupiter and Saturn, respectively, were launched on 3 March 1972 and 6 April 1973. On 3 December 1973, more than a year and a half after leaving Earth, Pioneer 10 swept through its closest approach to Jupiter. Approaching its rendezvous with the planet, the spacecraft mapped and measured Jupiter's radiation belts

The robot explorer Sojourner rolls down Pathfinder's ramp, making measurements to determine the composition of Martian soil. Pathfinder was the second in NASA's Discovery program of low-cost missions with highly focused scientific goals. (Courtesy of the National Air and Space Museum, Smithsonian Institution)

and its turbulent magnetosphere, measured its temperatures, photographed and studied its moons, and sent back stunningly beautiful images of the planet itself, swathed in multicolored bands of cloud. Pioneer 11 made an even closer approach to Jupiter just a year later, on 2 December 1974, performing the same sort of survey that had been conducted by its predecessor before swinging off toward a rendezvous with the ringed planet Saturn and its moon Titan in September 1979.

Voyagers 1 and 2, a pair of even more sophisticated spacecraft, followed Pioneers' path to the outer planets. Launched on 20 August (Voyager 2) and 5 September 1977 (Voyager 1), the latter flew the swifter trajectory of the two and brushed past Jupiter in March 1979, four months before Voyager 2. Together the two interplanetary travelers made new discoveries related to the planet's complex magnetosphere and took detailed and remarkable closeup photos of the planet and its four major moons.

Arriving at Saturn in the fall of 1980 and the summer of 1981, respectively, the Voyagers studied the atmosphere of the planet, photographed its cloudy surface, investigated the famous rings, and raised the number of known Saturnian moons to fifteen. Voyager 1, its mission complete, continued on toward an eventual exit from the solar system. Like the plaques mounted on the Pioneer spacecraft, designed to show another civilization what we looked like and where we lived, the Voyagers carried a calling card: a gold-plated copper record. "The Sounds of Earth" contained two hours of sound, music, and digitized images of Earth.

Four years after its encounter with Saturn, Voyager 2 became the first spacecraft to visit Uranus. Based in part on Voyager data, planetary scientists now believe that the planet consists of a dense molten core about the size of Earth lying at the bottom of an ocean of water and ammonia some 6,000 feet (1,820 meters) deep. The atmosphere is composed primarily of hydrogen and helium, with small amounts of methane, ammonia, and water vapor. Voyager discovered ten new moons orbiting Uranus and gathered information that raised the number of rings surrounding the planet to eleven. Voyager 2 flew by Neptune in August 1989, then followed its companion out of the solar system.

The Galileo spacecraft, launched by the Shuttle Atlantis on 18 October 1989, is taking a long and energy-efficient path to Jupiter. An inertial, upper-stage, solid-propellant rocket motor attached to the craft drove the vehicle on the first part of a six-year voyage that involved picking up the required energy by means of a Venus-Earth gravity-assist trajectory. Galileo swept close by Venus in February 1990; it then swung around the Sun, observing two comets in the process; flew by Earth in December 1990; returned for another close swing by Earth in December 1992; and then whipped toward Jupiter. During the process of building velocity assisted by gravity, Galileo's instruments returned scientific data on Earth, our Moon, Venus, and the comets Austin and Levy, but the failure of a communications antenna made matters difficult for ground controllers.

Galileo reached Jupiter in 1996. A probe launched by the spacecraft returned new information on the composition and nature of the Jovian atmosphere. In addition, Galileo helped scientists draw important new conclusions with regard to Jupiter's moons. There were three-dimensional images of the fractured, icy surface of Ganymede and new evidence that Jupiter's largest

Technicians work on the extended magnetometer boom during assembly of the Galileo spacecraft prior to its launch in 1989. In flight the boom, which measures 35.8 feet (10.9 meters), extended outward to isolate the magnetic sensors from the electrical fields of other instruments and the craft itself. (Courtesy of the Jet Propulsion Laboratory)

moon has a magnetic field. Information streaming back from Galileo suggested that "warm ice or even liquid water" might still exist beneath the frozen crust of Europa. Io, one of the most volcanically active objects in the solar system, had altered considerably since the stunning images returned by Voyager seventeen years before.

The outer planets lay beyond the reach of the Soviets, whose technology was less sophisticated than that of the United States. In 1988 the USSR undertook its most ambitious planetary mission to date. Two Phobos spacecraft were readied to orbit Mars at an altitude of only 500 miles (805 kilometers) and actually to land probes on the surface of the moon Phobos. Two months after launch, however, contact with the first spacecraft was lost as a result of a simple error in the instructions broadcast by ground controllers. Phobos 2 entered Mars orbit on 29 January 1989 and succeeded in returning some data. It was clear, however, that the spacecraft was experiencing problems. Ground controllers ordered the probe to change its orientation on 27 March and immediately lost contact with the craft. Phobos 2 had met the fate of its twin.

Roald Sagdeev, director of Russia's Institute for Space Research, could no longer contain himself. He laid the blame for the failure of the project squarely on the shoulders of the engineer-administrators who had ran the Soviet space program. In his view these administrators had compromised the scientific goals of the program for thirty years and violated the rules of common sense and good spacecraft design in an effort to rush forward at breakneck speed to satisfy their political masters. In the case of Phobos, coordination between the scientific and engineering teams had been poor. As usual the engineers had won the inevitable turf battles. "I hope that in future," Sagdeev remarked, "space technology producers [engineer-administrators] will have their absolute freedom restricted so that the world scientific community, as the end user of this technology, can have a say in making decisions on spacecraft design."[3]

U.S. spacecraft-design practice and communications protocols would have prevented the sequence of events that led to the loss of Phobos 1 and Phobos 2. Still, U.S. space scientists have echoed Sagdeev's point of view. NASA's great successes with automated scientific spacecraft obscure a fundamental tension within the organization. Since the early 1960s the scientists and engineers involved in the effort have often felt like second-class citizens of an agency whose highest priority is not the pursuit of space science.

This problem is reflected in the agency's organizational structure. Under T. Keith Glennan, all operational space activity, manned and unmanned, was the business of the Office of Spaceflight Programs. Soon after taking over the agency James Webb created a new Office of Manned Spaceflight (later renamed the Office of Space Transportation Systems, and now called the Office of Spaceflight) to manage astronaut activity. Instrumented space programs became the responsibility of an Office of Space Science and Technology. With the segregation of the total space effort into manned and unmanned programs, the proportion of the budget available for science and applications was drastically reduced. It was the beginning of a rift that continues within the agency to the present day. The lion's share of the funding has gone to support the hugely expensive and dramatic manned programs—Mercury, Gemini, Apollo, the Space Shuttle, and, most recently, the space station. Important scientific missions, such as the exploration of Halley's comet in 1985–86, at the time of its return to Earth, were abandoned for lack of funding.

The advent of the Space Shuttle, initially billed as a "space truck" specifically designed to carry satellite payloads into orbit, promised to draw the two segments of the total U.S. program back together. Instead, launch delays and a halt of Shuttle flights following the Challenger disaster threatened the survival of such scientific spacecraft as the Hubble Space Telescope and Galileo.

Scientists involved with satellite and planetary-probe flights point out that their effort has resulted in most of the great discoveries of the space age. Remote-sensing scientists have pioneered applications programs aimed at improving life on Earth. In fact, they have been most responsible for advancing critical technologies. "One of the sad things about the space shuttle and the space station," Burton I. Edelson, a veteran of Comsat and one-time head of NASA's Office of Space Science and Applications, once remarked, "is that they really are not technology drivers." Edelson underscored the achievements of the Voyager grand tour of the outer planets: "The two craft spent eight and one half years traveling to the far reaches of the solar system; they were reprogrammed and reoriented many times; and they could still point their cameras with enormous precision and return stunning images of superb quality on demand . . . *a billion miles away* . . . *after eight and a half years!*"[4]

Some space scientists, such as pioneer James Van Allen, have consistently favored a reduction of the manned space programs in favor of robot explorers. Others have found it politically expedient to offer support to the higher-visibility manned projects in order to ensure themselves a share of the pie. "You can't go in there to Congress and say, 'cut this manned crap and we could do some good science at a tenth the cost,'" Columbia University astrophysicist David Helfand has pointed out. "That's a disaster on Capitol Hill. You're co-opted into playing the game, into not criticizing any aspect of the [NASA] program."[5]

Another one of those great moments in the history of space flight arrived in 1996 and, as in the case of Pioneer 10, virtually no one noticed. For the first time the direct revenues resulting from commercial space operations rose above the total of approximately $77 billion expended on space by the governments of the world. Space was no longer the sole domain of military officers, bureaucrats, and astronauts. Businessmen are flourishing in orbit, and business will only improve with the passage of time.

It may seem that the question of an appropriate balance between manned space-flight programs and robot scientific spacecraft will not be fully resolved in the immediate future. In fact, more than four decades of space operations have largely settled the matter. However inspirational the sight of human footprints on the Moon, the real revolution of the space age is to be found in the achievements of robot spacecraft and their impact on human life and science.

18 • Outward Bound

There can be no thought of finishing, for "aiming at the stars," both literally and figuratively, is a problem to occupy generations, so that no matter how much progress one makes, there is always the thrill of just beginning. *Robert H. Goddard to H. G. Wells, 1932*

"No airship will ever fly from New York to Paris," Wilbur Wright remarked to an Illinois reporter in 1909. "That seems to me to be impossible." The engine was the problem. "No known motor can run at the requisite speed for four days without stopping," he explained, "and you can't be sure of finding the proper winds for soaring." Nor did he hold out any great hope for improved carrying capacity. "The airship will always be a special messenger," he predicted, "never a load-carrier." His brother Orville agreed, explaining to a Dayton reporter that he did not "believe that the airplane will ever take the place of trains or steamships for the carrying of passengers."[1]

Prognostication can be a difficult business, particularly in an era of rapid change. The inability of two brilliant inventors to forecast even the near-term progress of an infant technology to which they had given birth is not so difficult to understand. In 1909 the airplane had nowhere to go but up, and much faster, at that. Aeronautical progress was occurring at such a rapid rate that the world's first military airplane, the aircraft the Wright brothers had sold to the

U.S. Army in 1909, was judged to be obsolete and presented to the Smithsonian Institution as a historic object less than two years after it had entered service. At such a time choices abound, and the direction that a technology will move toward maturity may be far from certain.

Still, some elements of the near-term future seem fairly certain. If human beings are to continue to fly into space, for example, an orbiting station that can serve as a laboratory, a reconnaissance platform, a base for orbital operations, and a jumping-off point for journeys into deep space is a logical next step. Unless something goes badly wrong, the permanent human habitation of space will begin when the first crew takes up residence aboard the International Space Station.

With phase 1 (the Shuttle-Mir project) behind them, ISS planners began preparing for November 1998, when a Russian Proton booster would lift the very first element of the station, the large Zarya cargo block, into Earth orbit. NASA would follow a month later with the Unity node and two pressurized mating adapters. The first crew, astronaut Bill Shepherd and cosmonauts Yuri Gidzenko and Sergei Krikalev, were scheduled to take up residence aboard the ISS in July 1999. So it would go until the final flight in January 2004, when NASA would launch the final element, the U.S. habitation module.

There were problems, however. The traditional rivalry between space scientists and the engineers and administrators focused on a human presence in space was alive and well and living at NASA. Robert Park of the American Physical Society announced that the impact on science of the ISS plans "will be very, very slight," adding that "it is hard to imagine how we could spend our dollars on research in a less efficient way."[2] By March 1998 projected cost overruns on the part of Boeing, the U.S. prime contractor, and what NASA termed "newly recognized needs," had raised projected total costs from the agreed upon $17.4 billion to $20.6 billion. The *Wall Street Journal* reported that costs could ultimately rise as high as $24 billion, with the project's schedule stretching three years longer than currently planned.

For their part, the Russians were struggling not to fall behind, and, when they did, to put things in the best possible light. "The ISS project is the only way for Russia to continue a manned space program," Sergei Gorbunov, a spokesman for the Russian Space Agency, noted. He assured the United States that "the Russian government crisis will not prevent our participation in the project" and that all things appeared to be on time and on budget.[3]

U.S. officials were far less certain. In August 1998 NASA announced that it was fitting the Space Shuttle fleet with modified steering thrusters that would enable them to replace the Russian Soyuz spacecraft on at least some of the forty missions during which Soyuz was scheduled to resupply the growing station and would provide the occasional boost required to keep it in orbit. "With the first components of the international space station just over 100 days away from launch," the BBC reported, "NASA's decision will hurt the Russian pride."[4]

Given ever more serious economic and political problems, it is not certain what role the Russians will ultimately play in the ISS program. Moreover, faltering Asian economies could have an impact on the ability of other NASA partners to contribute to the space station. The current schedule could certainly be extended, however, it seems unlikely that NASA will abandon the goal of creating a permanent habitation in space.

The ISS has inspired research and development efforts on a new generation of spacecraft. The process began in 1994, when President Bill Clinton directed NASA to begin development of future reusable launch systems. The agency established the Space Transportation Program Office, the ultimate goal of which was to support the development of an efficient, reusable launch system that would reduce the per pound the cost of boosting a payload into low Earth orbit from ten thousand to one thousand dollars.

In an effort to draw private corporations into the effort, NASA and its industry partner, McDonnell Douglas, produced the DC-XA. The forty-three-foot tall, rocket-propelled, remotely piloted, vertical takeoff and landing technology demonstrator was designed as the first step toward the development of an operational single-stage-to-orbit (SSTO) reusable launch vehicle (RLV), a "space liner" capable of shuttling back and forth between Earth bases and the space station. The DC-XA completed three successful flights at the White Sands Missile Range in the summer of 1996. At the end of the fourth flight, however, the craft toppled over and exploded when one of the four landing gear failed to deploy. Less than a month after the accident, NASA signed a contract with the Orbital Sciences Corporation for the design, development, and testing of the X-34, a "technological bridge" between the DC-XA and the X-33, the first full-scale experimental single-stage-to-orbit RLV. The USAF has announced plans for a military space plane (MSP) said to be "complementary" to the NASA RLV.

NASA is also testing the prototype X-38 crew return vehicle (CRV). Based on lifting body technology, the CRV will be permanently attached to the In-

A prototype of the DC-XA single-stage-to-orbit reusable launch vehicle lifts off during a test at the White Sands Missile Range, New Mexico, in April 1995. The goal of the development program is to significantly reduce the cost of space launches. (Courtesy of NASA)

ternational Space Station, serving as an emergency "lifeboat" in which the crew can return to Earth. Once the basic emergency function has been demonstrated, NASA officials suggest that the CRV design might be modified to serve as "a sort of pick up truck" for the space station, capable of transporting crew members on short flights in the neighborhood of the orbiting base.

Other nations are also considering the development of spacecraft capable of carrying human beings into orbit. In October 1995 the Council of the European Space Agency approved a two-year study for a crew transport vehicle (CTV) that would provide ESA with a capacity to travel back and forth from orbit with a crew on board. The following spring ESA agreed to participate with NASA in a study of lifting body technology that might be shared by both the CRV and the CTV. That partnership continues, and the ESA has awarded a prime contract for a definition study of the CTV to a French, German, and Italian consortium.

Japan, another NASA partner with an active space program of its own, is also hoping to develop the capability to travel to and from orbit. That country's spacecraft, Hope-X, a winged, reusable, remotely controlled shuttle, would serve the needs of the International Space Station and allow Japan to expand its experience in a wide range of technologies. The launch of experimental vehicles designed to test systems and materials for Hope-X has been underway since early 1995. Japan has announced plans to launch Hope-X from the Tanegashima Space Center aboard an H-2A booster sometime in the year 2000. Operational flights with the Hope spacecraft would follow early in the twenty-first century.

Ultimately, China, most certainly not a partner in the ISS venture, also hopes to send its citizens into space. As early as 1979 a Shanghai newspaper published a photograph of a Chinese astronaut in training. At the end of the 1980s China unveiled an ambitious program that included both a mini-shuttle and a mini–space station. At an October 1996 meeting of the International Astronautical Federation in Beijing, officials went so far as to announce their intention to launch a Chinese citizen into space by the fiftieth anniversary of the nation in 1999.

Nor is the dream of space flight limited to the nation states of the world. The announcement, on 18 May 1996, of an X Prize to encourage the private development of a spacecraft captured the attention of both the aerospace community and the general public. Peter Diamandis, chairman and president of the St. Louis–based foundation offering the award explained that the purpose of the X Prize was to "stimulate the development of commercial space tourism by awarding a $10 million prize to the first team to build and fly a reusable spaceship capable of carrying three individuals on a sub-orbital flight."[5]

Inspired by the prizes that were so successful in stimulating the progress of aviation during the early years of the twentieth century, the X Prize has succeeded in inspiring preliminary designs on the part of several engineers and companies, including Burt Rutan, one of the most imaginative aeronautical engineers of his generation, and the man who designed the first airplane to fly around the world nonstop and unrefueled.

In January 1996 the Federal Aviation Administration's Office of Commercial Space Transportation reported that some two dozen RLV programs, both public and private, were underway around the globe. "Unwilling to wait for NASA's [RLV] program to validate new technologies required to develop a completely new type of spacecraft," the FAA report comments, "these companies are focusing on using existing technologies, materials, and components to build their own vehicles." And the market? "Some of these companies are also focused on space tourism and other potential new markets," the report concludes, "instead of, or in addition to, offering payload launch services."[6]

Not all of the emphasis will be on the development of high-technology reusable spacecraft in the immediate future. Conventional "throwaway" launch vehicles made a modest come back in the United States during the early 1990s. The USAF has contracted for the expendable boosters with which to launch its military satellites. Moreover, pressure from international competitors has encouraged U.S. civilian space planners to consider innovative and inexpensive approaches to providing launch services.

The Pegasus small-launch system, developed by the Orbital Sciences Corporation, has attracted a great deal of attention. Launched from beneath the wing of a flying airplane, the Pegasus rocket made its first flight to orbit on 5 April 1990 and promises to reduce drastically the cost of boosting relatively lightweight payloads into near-Earth space.

What does the future hold? It is clear that the world of Stanley Kubrick's film, *2001: A Space Odyssey,* with its luxuriously appointed space station, Pan American space liners, sophisticated lunar bases, and human expeditions to Jupiter will not arrive on schedule. Still, a more modest orbital base, and a new generation of reusable spacecraft, are reasonably close at hand. The shape of the more distant future is less certain.

Since the 1970s any number of special-interest groups have been willing to suggest appropriate directions for our long-term future in space. Some organizations were dedicated to the realization of a particular idea. The L-5 Society sought to gain support for the notion, popularized by physicist Gerard K. O'Neil, of establishing large space colonies between Earth and the Moon. Others, such as the Planetary Society, seek to generate public enthusiasm for space-science programs. A newcomer on the scene, the Mars Society, is dedicated to sponsoring a privately financed mission to Mars.

But NASA will be expected to realize its dreams, a fact that adds considerably to the difficulty of the planning process. In 1987, eager to incorporate projects such as the space station into a integrated blueprint for the future of the American space effort, NASA administrator James Fletcher asked Sally Ride, the first U.S. woman to fly into space and a member of the Rogers Commission, to head a task group that would suggest appropriate directions for the agency. In her subsequent report, *Leadership and America's Future in Space,* Ride noted that the U.S. civilian space program stood at a crossroads. "Two fundamental, potentially inconsistent views have emerged," she noted. "Many people believe that NASA should adopt a major, visionary goal. They argue that this would galvanize support, focus NASA programs, and generate excitement. Many others believe that NASA is already overcommitted in the 1990s; they argue that the space agency will be struggling to operate the Space Shuttle and build the Space Station, and could not handle another major program."[7]

Pointing out that the United States no longer can hope to maintain leadership in every area of space flight, the members of the task group suggested that NASA pursue four basic initiatives:

1. Mission to planet Earth. With the space station and nine specialized satellites in orbit by the year 2000, NASA and cooperating international partners could keep a careful watch over global cloud cover, changes in the earth's vegetation, alterations in the polar caps, worldwide rainfall and moisture, the chemical composition of the oceans, shifts in the earth's tectonic plates, and changes in the structure and composition of the atmosphere.
2. Exploration of the solar system. NASA should proceed with the long-

delayed Magellan and Galileo missions in 1989–93. In addition, several new missions should be added to the current program. A comet rendezvous asteroid flyby (CRAF) should be launched in 1993, and three robot missions to Mars should be dispatched from the space station between 1996 and 1999. The Cassini spacecraft should begin exploring Saturn and its moon Titan in 2005.

3. Outpost on the Moon. During the 1990s a sophisticated, unmanned reconnaissance of the Moon could be undertaken, the object of which would be to locate a site for the first permanent lunar base. If the first construction crew arrived on the Moon sometime between 2000 and 2005, by the year 2010 the base would be able to support thirty human beings for months at a time.

4. Humans to Mars. Human beings should establish a base on Mars during the second decade of the twenty-first century. Their spacecraft could be assembled in Earth orbit, with the space station serving as a construction base. "A successful Mars initiative," the report concluded, "would recapture the high ground of world space leadership and would provide an exciting focus for creativity, motivation, and pride of the American people. The challenge is compelling, and it is enormous"[8]

It was a very tall order. Some elements of the Ride report were acted upon immediately. Magellan did fly to Venus, and Galileo was put back on track. Just as obviously, the more visionary recommendations could not be achieved without a national commitment of the sort that sent astronauts to the Moon.

President George Bush did his best to provide that commitment. On 20 July 1989—the twentieth anniversary of the first Moon landing—he announced the Space Exploration Initiative (SEI), a thirty-year plan embodying the major recommendations of the Ride report. SEI called for a permanent human presence on the Moon, to be followed by an expedition to Mars before the year 2020. Whereas John Kennedy's challenge to fly to the Moon had been met with overwhelming public enthusiasm and wholehearted support from Congress, George Bush's SEI proposal scarcely drew a murmur from the public. More important, Congress refused to fund the first year of work on the longer-range aspects of the program. NASA administrators recognized that they would be fortunate just to preserve the funding for a basic space station.

In 1990 the National Space Council, a presidential advisory board over-

John H. Glenn prepares for his return to space in 1998, thirty-six years after his Mercury flight. The seventy-seven-year-old former Ohio senator was tested to determine the effects of space travel on an older body. NASA believes the tests will also benefit older persons here on Earth. (Courtesy of NASA)

seeing matters relating to space policy, ordered a planning assessment of its own. Norman Augustine, chairman and chief executive officer of the Martin Marietta Corporation, was asked to head a twelve-member advisory committee on the future of the U.S. space program. The Augustine panel considered current NASA plans for the future and offered policy recommendations for a rational, long-term space effort.

The committee's study, released in December 1990, was interpreted in a variety of ways. It gave strong support to the near-term aspects of the Ride report and to the SEI, calling for a renewed emphasis on space-science satellites and probes and on programs related to space-based environmental research. With regard to manned space initiatives, however, the Augustine study was a far cry from the Ride report.

The Augustine committee suggested that every effort be made to encourage the implementation of an advanced launch system of expendable boosters that would be developed jointly by the USAF and NASA. The Shuttle should be used only when a human presence in space is absolutely necessary, it said. NASA's request for a new sixth Shuttle was seen as ill advised. Experiments conducted aboard a space station should concentrate on the study of human physiology in space—not on preparations for missions to the Moon or to Mars. The Augustine report recognized a basic and inescapable truth about the near future of the American space effort: Barring the return of another era of overwhelming political and public support for space flight, NASA would have to proceed slowly and cautiously. The agency would have its hands full flying the Shuttle, building the space station, and maintaining its unmanned applications and space-science programs through the turn of the century.

But if there is reason for short-term caution, the longer-range view is at least mildly optimistic. Exploration has been a hallmark of Western civilization, with both geographic expansion and the discovery and exploitation of new resources the keys to its success. It is part of our very nature to wonder what lies over the next hill—and to want to possess it. Whatever the cost, it is unlikely that human beings will be able to resist the challenge of the solar system.

Common sense and limited budgets guarantee that robot explorers will continue to lead the way in space exploration. Someday, probably in the not-too-distant future, human beings will establish a permanent habitation on the Moon. We will not only visit Mars but also probably live there for extended periods of time. Ultimately, we will stand on all the other points in the solar system on which it is possible to do so. Jupiter and Saturn may not have a solid surface, but their moons do. It is unlikely that the government of the United States, or of any other nation, will be willing or able to bear the full cost of such ventures. International cooperation on a variety of levels already is an important feature of both manned and unmanned space operations, and there is every reason to believe that will remain the case in the future.

Human voyages to the stars are another matter. Technological breakthroughs and new discoveries notwithstanding, the velocity of light seems to

Payload commander Jan Davis *(left)* and mission specialist Robert Curbeam prepare for a future Space Shuttle mission, July 1997. Despite reduced budgets and diminished public interest, space flight will continue to fascinate those among us who have always wondered what lies just over the next hill. (Courtesy of NASA)

represent a final cosmic speed limit. Either there are some time/space short-cuts we have yet to discover, or we may have to resort to the use of ships on which generations of crew members would be born, grow old, and die wandering across a relatively small section of the immediate interstellar neighborhood in search of a habitable planet. Surely something more than the simple desire to see over the next hill would be required to impel us on such a journey. Nor would we likely undertake an interstellar voyage in search of other beings unless we knew precisely where they were located. What reasons might lead us to move beyond the solar system?

It was a question that fascinated Italian journalist Oriana Fallaci, who toured NASA centers during the early years of the Apollo program, interviewing everyone from astronauts to Wernher von Braun. Everywhere she went, she posed one simple question: "Why should the human race seek to leave its home planet?" The answer came in a variety of ways, but its essence was always the same: "Well, if the sun ever dies . . . "[9]

Konstantin Tsiolkovsky and Robert Goddard had responded to that question in the same way. A sense of cosmic destiny, a need to arrange for the ultimate survival of the species, was a key element of the obsession that bound the lives of these two men. Tsiolkovsky offered his thoughts on the evolution of space flight in a paper published in 1926. The colonization of space would begin with the construction of permanent space stations in Earth orbit. The stations would tap solar energy to meet the needs of Earth and to propel other spacecraft to the planets. Industry based on resources obtained from asteroids and other "small bodies" in the solar system would flourish in space. Eventually, Tsiolkovsky predicted, "the population of the solar system will become one hundred thousand million times greater than the present population of the Earth. A limit will be reached, after which it will be necessary to emigrate over the entire Milky Way." Finally, "the sun will start to die. The remaining population of the solar system will move to other stars to join their brothers who departed earlier."[10]

Robert Goddard was as secretive about his ultimate justification for space flight as he was about every other aspect of his work. On Monday, 14 January 1918, the thirty-five-year-old physicist prepared a short essay on the subject, entitled "The Last Migration." Across the face of the envelope into which he tucked the document, he wrote a set of instructions: "To be given to the Smithsonian Institution, after the owner has finished with it, there to be preserved on file, and used at the discretion of the Institution. The notes should be read thoroughly only by an optimist."[11]

In the essay Goddard asked his "optimistic" reader to imagine a time in the far distant future when the Sun has grown dim and cool. One by one, great interstellar "arks," powered by "intra-atomic energy" and filled with people and libraries containing the sum of all human knowledge, are dispatched to the far corners of the Milky Way. Once such a craft is well on its way, it will be allowed to cool "to the temperature of space." The contents of the ship would "remain motionless and inanimate throughout. . . . After long intervals of time," a special device, perhaps a "radioactive alarm clock," would begin to warm and reanimate the crew. Surviving those conditions, Goddard recognized, would present something of a problem. "It has long been known," he pointed out, "that protoplasm can remain inanimate for great periods of time, and can also withstand great cold, if in the granular state." Perhaps it would be necessary "to evolve suitable beings through many generations." Or perhaps granular protoplasm itself, "suitably enclosed, might be sent out of the solar system, this protoplasm being of such a nature as to produce human beings, in time, by evolution."[12]

Space flight is about leaving. It is a daunting prospect. For all our knowledge of the cosmos, we can identify only one planet that nurtures life. There may be other havens out there, but finding them will not be easy. Forty years of space exploration has taught us that our own solar system is far less hospitable than we once hoped.

Even as late as the 1940s some astronomers still held that the clouds of Venus sheltered verdant, carboniferous swamps, and that a practiced eye could discern delicate patches of green spreading across the surface of Mars as the planet's polar ice caps melted each spring. The Mariner, Venera, Vega, and Magellan spacecraft have demonstrated that pressure, temperature, and a noxious atmosphere combine to make the surface of Venus a place where specially armored probes survive for only a few minutes. Vikings 1 and 2 surprised all but the most pessimistic researchers by failing to find the slightest indications of life on the Red Planet. In all the solar system but Earth, we have yet to find a single drop of free water.

Some scientists suggest that we can change all that through the practice of "terra-forming," building an atmosphere and transforming an apparently barren world like Mars into a planet that can support life. Given our inability to understand or manage the complexity of our native environment, that sounds like an open invitation to ecological disaster.

We are left with the fact that space is relentlessly hostile, and we human beings are extraordinarily frail. Only by shrouding ourselves in elaborate machinery can we venture beyond our world. Still, we can reasonably define ourselves as creatures whose capabilities of brain and hand enable us to achieve that which is physically possible. That being the case, and presupposing the continued survival of our species, it is fair to assume that some day human beings will set out on the ultimate journey to the stars.

Perhaps as Robert Goddard suggested, the voyage will be a "last migration" from a dying Sun. Perhaps it will be impelled by a simple desire to stand before the members of another species with whom we have been in contact for generations. Whatever the reason, the words of Johannes Kepler, the man who first understood the basic mechanism of the solar system, remain as true today as they were three and one half centuries ago: "Provide ship or sails adapted to the heavenly breezes, and there will be some who will not fear even that void."

Notes

1. A Plurality of Worlds

1. Carola Baumgart, *Johannes Kepler: Life and Letters* (New York: Philosophical Library, 1951), 17.
2. Galileo Galilei quoted in Arthur Koestler, *The Sleepwalkers: A History of Man's Changing Vision of the Universe* (New York: Grosset & Dunlap), 356.
3. I. Bernard Cohen, "An Interview with Einstein," in *Einstein: A Centenary Volume,* ed. A. P. French (Cambridge: Harvard University Press, 1979), 41.
4. Johannes Kepler quoted in Koestler, *Sleepwalkers,* 373.

2. The Call of the Cosmos

1. R. H. Goddard, "Material for an Autobiography of R. H. Goddard," in *The Papers of Robert H. Goddard,* ed. Esther C. Goddard and G. Edward Pendray (New York: McGraw-Hill, 1970), 1:9 (hereafter cited as *Papers of Robert H. Goddard*).
2. R. H. Goddard to H. G. Wells, 20 April 1932, *Papers of Robert H. Goddard* 2: 821–22.
3. R. H. Goddard Diary, 10 November 1938, *Papers of Robert H. Goddard* 3:1216.
4. R. H. Goddard Diary, 19 October 1939, *Papers of Robert H. Goddard* 3:1277.
5. K. E. Tsiolkovskii, "Autobiography," in *The Coming of the Space Age,* ed. Arthur C. Clarke (New York: Meredith Press, 1967), 100.
6. N. A. Rynin, *Interplanetary Flight and Communications,* NASA TT F-646, vol. 3, no. 7, in *K. E. Tsiolkovskii: Life, Writings, and Rockets,* by N. A. Rynin (Jerusalem: Israel Program for Scientific Translations, 1971), 29.
7. Ibid., 29.
8. N. A. Rynin, "The Autobiography of K. E. Tsiolkovskii," in Rynin, *K. E. Tsiolkovskii,* 3.
9. Rynin, "Autobiography of K. E. Tsiolkovskii," 3.
10. K. E. Tsiolkovskii, *Exploration of the Universe with Reactive Devices,* in *Collected*

Works of K. E. Tsiolkovskii, ed. B. N. Yur'yev, vol. 2, *Reactive Flying Machines,* NASA TT F-237 (Washington, D.C.: NASA, 1965), 213.

11. R. H. Goddard Diary, 8 August 1915, *Papers of Robert H. Goddard* 1:163.

12. R. H. Goddard Diary, 15 October 1938, *Papers of Robert H. Goddard* 3:1209.

13. R. H. Goddard, "A Method of Reaching Extreme Altitudes," *Smithsonian Miscellaneous Collections* 71, no. 2, Washington, D.C., 1919.

14. Bronx Exposition, Inc., to Smithsonian Institution, 12 January 1920, *Papers of Robert H. Goddard* 1:408–9.

15. Hermann Oberth, "Autobiography," in *The Coming of the Space Age,* ed. Arthur C. Clarke (New York: Meredith Press, 1967), 114; see also Hermann Oberth, "My Contributions to Astronautics," in *First Steps toward Space,* ed. F. C. Durant III and George S. James (Washington, D.C.: Smithsonian Institution Press, 1974), 129–40.

16. R. H. Goddard to Secretary, Smithsonian Institution, 1 August 1923, *Papers of Robert H. Goddard* 1:423.

17. R. H. Goddard Diary, 6 December 1925, *Papers of Robert H. Goddard* 2:574.

3. Raketenrummel

1. Robert H. Goddard Diary, 16–17 March 1926, *Papers of Robert H. Goddard* 2: 580–81.

2. Herbert Lehman, *This High Man: The Life of Robert Goddard* (New York: Fararr, Strauss & Giroux, 1963), 129.

3. A wide selection of Goddard "fan mail" from various periods is included in all three volumes of the *Papers of Robert H. Goddard.* See also Lehman, *This High Man,* 102–3.

4. L. K. Korneev, ed., *Tsander: Problems of Flight by Jet Propulsion,* NASA TT-147 (Jerusalem: Israel Program for Scientific Translations, 1964), 15–16.

5. Willy Ley, *Rockets, Missiles and Men in Space* (New York: Viking Press, 1968), 105.

6. Willy Ley, "Count von Braun," *Journal of the British Interplanetary Society* 6 (June 1947): 155.

7. Dimitri Marianoff and Palma Wayne, *Einstein: An Intimate Study of a Great Man* (New York: Doubleday, Doran, 1944), 115, cited in Frank Winter, *Prelude to the Space Age: The Rocket Societies, 1924–1940* (Washington, D.C.: Smithsonian Institution Press, 1983). Winter's book is a detailed guide to the subject of the rocket societies.

8. Rudolph Nebel, *Rakentenflug,* NASA TT F-11173, August 1967 (Berlin, 1932), 8.

9. Walter Dornberger, *V-2* (New York: Viking Press, 1954), 33.

4. An American Dreamer

1. G. E. Pendray, "Recent Worldwide Advances in Rocketry," *Bulletin of the American Interplanetary Society* 14 (December 1931): 1, is an abstract of the original lecture.

2. For one solid account of Hugo Gernsback and the early history of pulp "scientific-tion" in the United States, see Sam Moskowitz, *The Immortal Storm: A History of Science Fiction Fandom* (Westport, Conn.: Hyperion Press, 1974), 4–7; Brian W. Aldiss, *Billion Year Sprees: The True History of Science Fiction* (New York:

Schocken Books, 1974), 209 – 16. This standard history of the science fiction genre also provides information on the career of Hugo Gernsback.

3. R. H. Goddard to G. E. Pendray, *Bulletin of the American Interplanetary Society* 10 (June–July 1931): 9 – 10.

4. Charles A. Lindbergh, *Autobiography of Values* (New York: Harcourt Brace Jovanovich, 1976), 337.

5. Ibid., 339.

6. R. H. Goddard to C. G. Abbot, 29 November 1929, *Papers of Robert H. Goddard* 2:714.

7. Lindbergh, *Autobiography of Values,* 340.

8. Ibid., 341.

9. R. H. Goddard to R. A. Millikan, 1 September 1936, *Papers of Robert H. Goddard* 2:1012 – 13.

10. Frank J. Malina, "On the GALCIT Rocket Research Project, 1936 – 1938," in *First Steps toward Space,* ed. Frederick C. Durant III and George S. James (Washington, D.C.: Smithsonian Institution Press, 1974), 137.

11. R. H. Goddard to Harry F. Guggenheim, 15 December 1944, *Papers of Robert H. Goddard* 3:1553.

12. "Germany's Flying Vengeance Bomb V-2 Follows American Pre-War Rocket Design," *National Geographic News Bulletin,* 19 January 1945, Washington, D.C.

13. Theodore von Kármán and Lee Edson, *The Wind and Beyond* (Boston: Little, Brown, 1967), 240 – 42.

14. Lindbergh, *Autobiography of Values,* 400 – 401.

5. Vergeltungswaffe!

1. Frederick Schneikert, in *The Rocket Team: From the V-2 to the Saturn Moon Rocket,* ed. Frederick I. Ordway and Mitchell R. Sharpe (New York: Thomas Y. Crowell, 1979), 2–3.

2. Frank Winter, *Rockets into Space* (Cambridge: Harvard University Press, 1990), 46.

3. "A Reporter at Large: A Romantic Urge," *New Yorker,* 21 April 1951, 77. This was von Braun's favorite justification for involvement with the Nazi military. See a similar comment by von Braun in Shirley Thomas, *Men of Space* (Philadelphia: Chilton, 1960), 137.

4. Ley, "Count von Braun," 154.

5. Ibid.

6. Wernher von Braun, "Reminiscences of German Rocketry," *Journal of the British Interplanetary Society* 15, no. 3 (May–June 1956): 125–45.

7. Walter Dornberger, "The First V-2," in *The Coming of the Space Age,* ed. Arthur C. Clarke (New York: Meredith Press, 1967), 29–30.

8. Michael Neufeld, *The Rocket and the Reich: Peenemünde and the Coming of the Ballistic Missile Era* (New York: Free Press, 1995), 273.

9. Ordway and Sharpe, *Rocket Team,* 248–49.

10. Linda Hunt, *Secret Agenda: The United States Government, Nazi Scientists, and Project Paperclip* (New York: St. Martin's Press, 1991), 64.

11. Ibid.

12. Ibid., 18.
13. Lindbergh, *Autobiography of Values,* 348–49.
14. Dornberger, "First V-2," 30.
15. Ordway and Sharpe, *Rocket Team,* 22.
16. Thomas Franklin, *An American in Exile: The Story of Arthur Rudolph* (Huntsville, Ala.: Christopher Kaylor, 1987), 42.
17. "Transcript of OSI Interrogation of Arthur Rudolph," in ibid., 332.
18. Tom Bower, *The Paperclip Conspiracy: The Hunt for the Nazi Scientists* (Boston: Little, Brown, 1987), 111.
19. Hunt, *Secret Agenda,* 226.
20. Neufeld, *Rocket and the Reich,* 187.

6. "Our Germans"
1. Malina, "On the GALCIT Rocket Research Project," 114.
2. Federal Bureau of Investigation, "John Whiteside Parsons, aka 'Jack' Parsons," 2 November 1950, Report of Interrogations and Investigation in the John Parsons biographical file, National Air and Space Museum Archive, Washington, D.C.
3. Frank J. Malina, "Origins and First Decade of the Jet Propulsion Laboratory," in *The History of Rocket Technology,* ed. Eugene Emme (Detroit: Wayne State University Press, 1964), 53–54.
4. Richard Porter, "Interview with David DeVorkin," 16 April 1984, 42. Transcript in the Department of Space History, National Air and Space Museum, Washington, D.C.
5. Dieter Huzel, *Peenemünde to Canaveral* (Englewood Cliffs, N.Y.: Prentice-Hall, 1962), xx.
6. Fritz Zwicky, *Report on Certain Phases of War Research in Germany* (Pasadena, Calif.: Aerojet Engineering, 1945), 238.
7. Porter, "Interview with David DeVorkin," 36.
8. Rabbi Stephan Wise, undated news article in scrapbook, Wernher von Braun Papers, Manuscript Division, Library of Congress.
9. "Warheads to Peaceheads," *Stars and Stripes;* Albert Deutsch, "Scientists Shocked by U.S. Efforts to Place Nazis in School Jobs Here," undated articles, scrapbook, Wernher von Braun Papers, Manuscript Division, Library of Congress.
10. United States Army Air Force, *Third Report of the Commanding General of the Army Air Forces to the Secretary of War,* by Gen. Henry H. Arnold, 12 November 1945, p. 68, quoted in R. Cargill Hall, "Origins of U.S. Space Policy: Eisenhower, Open Skies, and Freedom of Space," in *Exploring the Unknown: Selected Documents in the History of the U.S. Space Program,* ed. John Logsdon (Washington, D.C.: NASA, 1995), 1:213–15.
11. Project RAND, *Preliminary Design of a World-Circling Spaceship* (Santa Monica, Calif.: Project RAND, May 1926); for additional information on the report, see R. Cargill Hall, "Early U.S. Satellite Proposals," in *The History of Rocket Technology,* ed. Eugene Emme (Detroit: Wayne State University Press, 1964), 67–93.
12. Huzel, *Peenemünde to Canaveral,* 216.
13. Homer Newell, *Beyond the Atmosphere: Early Years of Space Science* (Washington, D.C.: NASA, 1980), 35.
14. Holger Toftoy, "Radio Script," in 30 September 1946, Toftoy Collection, Alabama

Space and Rocket Center, in David DeVorkin, "Warheads into Peaceheads: Holger N. Toftoy and the Public Image of the V-2 in the United States," *Journal of the British Interplanetary Society* 45 (1992): 442.

15. Frank J. Malina, "America's First Long-Range Missile and Space Exploration Program: The 'ORDCIT' Project of the Jet Propulsion Laboratory, 1943–1946: A Memoir," in *Essays on the History of Rocketry and Astronautics: The Proceedings of the Third through the Sixth History Symposia of the International Academy of Astronautics,* ed. R. Cargill Hall (Washington, D.C.: NASA, 1977), 339.

16. Ibid., 340–41.

7. Selling Space Flight

1. Arthur C. Clarke, *Astounding Days: A Science Fiction Autobiography* (New York: Bantam Books, 1989).

2. Tom Lehrer, "Wernher von Braun," *That Was the Week That Was,* phonograph album 6179-2, 1965.

3. G. A. Tokaty-Tokaev, in Michael Stoiko, *Soviet Rocketry: Past, Present, and Future* (New York: Holt, Rinehart and Winston, 1970), 74.

4. G. A. Tokaty-Tokaev, in *The Heavens and the Earth: A Political History of the Space Age,* ed. Walter A. McDougal (New York: Basic Books, 1985), 53.

5. James Harford, *Korolev: How One Man Masterminded the Soviet Drive to Beat America to the Moon* (New York: Wiley, 1997), 116.

6. Nikita Khrushchev, *Khrushchev Remembers: The Last Testament* (Boston: Little, Brown, 1974), 47.

7. Ibid.

8. Fellow Travelers

1. Yaroslav Golovanov, *Sergei Korolev: The Apprenticeship of a Space Pioneer* (Moscow: Mir Publishers, 1975), 7.

2. Ibid., 11.

3. Lyndon Baines Johnson, *The Vantage Point: Perspectives of the Presidency, 1963–1969* (New York: Holt, Rinehart and Winston, 1971), 272.

4. Maj. Gen. John B. Medaris, *Countdown for Decision* (New York: G. P. Putnam's Sons, 1960), 155.

5. Lynne Daniels, "Statement of Prominent Americans on the Opening of the Space Age," NASA Historical Note No. 22, NASA Headquarters.

6. Ordway and Sharpe, *Rocket Team,* 382.

7. McDougal, *Heavens and the Earth,* 175.

8. *Washington Post,* 21 November 1957.

9. Dwight D. Eisenhower, "Farewell Radio and Television Address to the American People," *Presidential Papers of President Dwight D. Eisenhower,* 1038–39.

9. This New Ocean

1. James Swenson Jr., James Grimewood, and Charles C. Alexander, eds., *This New Ocean: A History of Project Mercury* (Washington, D.C.: NASA, 1966), 284.

2. M. Scott Carpenter et al., *We Seven: By the Astronauts Themselves* (New York: Simon and Schuster, 1962), 7.

3. Ibid., 17.

4. Khrushchev, *Khrushchev Remembers*, 53.

5. Steven Zaloga, *Target America: The Soviet Union and the Strategic Arms Race* (Novato, Calif.: Presidio Press, 1993), 196.

6. Ibid., 196.

10. Racing to the Moon

1. Nikolai Tsymbal, ed., *First Man in Space: The Life and Achievement of Yuri Gagarin* (Moscow: Progress Publishers, 1984), 68.

2. McDougal, *Heavens and the Earth*, 247.

3. Ibid., 246.

4. John Fitzgerald Kennedy, "Urgent National Needs," in *Exploring the Unknown: Selected Documents in the History of the U.S. Space Program*, ed. John Logsdon (Washington, D.C.: NASA, 1995), 1:453.

5. Swenson, Grimewood, and Alexander, *This New Ocean*, 284.

6. "Report to the President-Elect of the Ad Hoc Committee on Space," in *Exploring the Unknown: Selected Documents in the History of the U.S. Space Program*, ed. John Logsdon (Washington, D.C.: NASA, 1995), 1:421.

7. Joseph J. Trenta, *Prescription for Disaster: From the Glory of Apollo to the Betrayal of the Shuttle* (New York: Crown, 1987), 3.

8. Roger Bilstein, *Stages of Saturn: A Technical History of the Apollo/Saturn Launch Vehicles* (Washington, D.C.: NASA, 1980), 63.

9. Carpenter et al., *We Seven*, 455.

10. Ibid., 335.

11. Michael Collins, *Liftoff: The Story of America's Adventure in Space* (New York: Grove Press, 1988), 59–60.

11. The Giant Leap: Gemini

1. Michael Collins, *Carrying the Fire: An Astronaut's Journeys* (New York: Ballantine Books, 1974), 1.

2. Ibid., 22.

3. Leonid Vladimirov, *The Russian Space Bluff* (New York: Dial Press, 1973), 108–9; Buzz Aldrin and Malcom McConnell, *Men from Earth* (New York: Bantam Books, 1989), 86.

4. Vladimirov, *Russian Space Bluff*, 143; James Oberg, *Red Star in Orbit* (New York: Random House, 1981), 80–81; Aldrin and McConnell, *Men from Earth*, 136.

5. Aldrin and McConnell, *Men from Earth*, 140.

6. Collins, *Liftoff*, 90.

12. The Apollo Era

1. Collins, *Liftoff*, 127.

2. Ibid., 127–28.

3. Wayne Biddle, "A Great New Enterprise," *Air & Space Smithsonian* 4, no. 7 (June/July 1989): 32–33.
4. Ibid., 32.
5. Ibid., 35.
6. Ibid., 37.
7. T. A. Heppenheimer, "Requiem for a Heavyweight," *Air & Space Smithsonian* 4, no. 7 (June/July 1989): 32–33.
8. Ibid., 52.
9. Ibid.
10. Ordway and Sharpe, *Rocket Team,* 398.
11. *Marshall Star,* 3 November 1965, in Bilstein, *Stages to Saturn,* 291.

13. Men from Earth

1. Biddle, "Great New Enterprise," 37.
2. Edgar Mitchell, "A Deep Encounter with the Cosmos," in *Footprints: The Twelve Men Who Walked on the Moon Reflect on Their Flights, Their Lives, and the Future,* ed. Douglas MacKinnon and Joseph Baldanza (Washington, D.C.: Acropolis Books, 1989), 91.
3. Neil Armstrong and Jacob Zent, quoted in ibid., 283–84.
4. Milton O. Thompson, *At the Edge of Space: The X-15 Flight Program* (Washington, D.C.: Smithsonian Institution Press, 1992), 15.
5. Collins, *Carrying the Fire,* 58.
6. Thompson, *At the Edge of Space,* 16.
7. Aldrin and McConnell, *Men from Earth,* 199–200.
8. Collins, *Carrying the Fire,* 59.
9. Collins, *Liftoff,* 7.
10. Charles Conrad Jr. and Alan B. Shepard, "Ocean of Storms and Frau Mauro," *Apollo Expeditions to the Moon* (Washington, D.C.: NASA, 1975), 226.
11. Harrison H. Schmitt, "The Great Voyages of Exploration," *Apollo Expeditions to the Moon* (Washington, D.C.: NASA, 1975), 271.
12. James A. Lovell, "Houston, We've Had a Problem," *Apollo Expeditions to the Moon* (Washington, D.C.: NASA, 1975), 262.

14. Salyut and Skylab: Rooms with a View

1. James Oberg and Alcestis Oberg, *Pioneering Space: Living on the Next Frontier* (New York: McGraw-Hill, 1986), 9.
2. Ibid., 8.
3. Ibid., 12.
4. Ibid.
5. W. David Compton and Charles D. Benson, *Living and Working in Space: A History of Skylab* (Washington, D.C.: NASA, 1983), 284.
6. Ibid., 308.
7. Aleksandr Aleksandrov, in Oberg and Oberg, *Pioneering Space,* 9.
8. Aleksandr Ivanchenkov, in ibid., 182–83.

9. Compton and Benson, *Living and Working in Space*, 307.
10. Ibid., 293.
11. Oberg and Oberg, *Pioneering Space*, 7.
12. Collins, *Liftoff*, 174.
13. Oberg and Oberg, *Pioneering Space*, 41.
14. Ibid.
15. Ibid., 45.
16. Ibid., 251.
17. Ibid., 248.
18. Ibid., 237.
19. Ibid., 248.

15. The Shuttle Era

1. Collins, *Liftoff*, 235.
2. Phillip Clark, *The Soviet Manned Space Program* (New York: Crown Books, 1988), 151.

16. At Home in Orbit

1. "Shannon's Letter Home," NASA Shuttle–Mir web site, http://www.shuttle.wip.nasa.gov/history/shuttle-mir/fidit.html (hereafter cited as NASA Shuttle–Mir web site).
2. Shannon W. Lucid, "Six Months on Mir," *Scientific American* (May 1998): 47.
3. "Mir Core Module," see NASA Shuttle–Mir web site.
4. Daniel Goldin, "Cooperation—Phase 1 Overview." See NASA Shuttle–Mir web site.
5. "Interview with John Blaha," NASA Shuttle–Mir web site.
6. "U.S. Astronaut Tells of Life on Russia's Mir," *New York Times*, 4 April 1995.
7. Lucid, "Six Months on Mir," 47.
8. "Pink Socks and Jello: Shannon Lucid Writes a Letter Home," NASA Shuttle–Mir web site.
9. 11 October 1996, "Interview with John Blaha."
10. "Shannon Lucid Post-Flight Press Conference," NASA Shuttle–Mir web site.
11. Jerry Linenger, "Letter to My Son" 29 March 1997, "Alone in a Dark Room," NASA Shuttle–Mir web site.
12. "Mir Floating from One Crisis to the Next," BBC News Online, http://www.news.bbc.co.uk/network/news/space.
13. "Mir Cosmonauts Lash out at Critics, Including Yeltsin," NASA Shuttle–Mir web site.
14. "Trouble in Orbit Sparks Tempest in Washington," CNN.com/TECH/space/9804/02/mir.retrospective/part3.html.
15. Ibid.
16. Frank Culbertson, quoted in "Mission Impossible: Culmination of Shuttle Mir Docking Program Provides a Measure of Confidence for International Space Station," NASA Shuttle–Mir web site.
17. "Mission Impossible," NASA Shuttle–Mir web site.

17. Robot Servants and Explorers

1. Arthur C. Clarke, "Everybody in Instant Touch," in *The Coming of the Space Age,* ed. Arthur C. Clarke (New York: Meredith Press, 1967), 147.
2. "Mars Pathfinder Winds Down after Phenomenal Mission," Public Information Office, JPL, CIT, 4 November 1997, http://www.jpl.nasa.gov/mpf-pressrel.html.
3. "Phobos Lost—Spacecraft Designers Blamed," *Spaceflight* (July 1989): 219; see also "Soviets See Little Hope of Controlling Spacecraft," *New York Times,* 10 September 1988.
4. William E. Burrows, *Exploring Space: Voyages in the Solar System and Beyond* (New York: Random House, 1990), 360.
5. Ibid., 364.

18. Outward Bound

1. Wilbur Wright, "Airship Safe: Air Motoring No More Dangerous than Land Motoring," *Cairo (Ill.) Bulletin,* 25 March 1909; "Catherine [*sic*] Wright Going Abroad to Witness Brother's Triumph," *Dayton Herald,* 2 January 1909.
2. "Space Station Countdown Underway," BBC News web site, http://www.news.bbc.co.uk/hi/english/sci/tech (hereafter cited as BBC News web site).
3. "Yet More Station Delays," http://www.ABCNews.com/sections/science/daily/news/space station.
4. "NASA Dumps Russia," BBC News web site.
5. "From the Chairman," hyperlink http://www.xprize.org/info/index.html.asp.
6. Quoted in Leonard David, "Space Tourism: Escape Velocity Vacations," *Final Frontier* (August 1998): 29–34.
7. Sally K. Ride, *Leadership and America's Future in Space: A Report to the Administrator* (Washington, D.C.: NASA, 1987), 5.
8. Ibid., 35.
9. Oriana Fallaci, *If the Sun Ever Dies* (New York: Atheneum, 1967).
10. "The Investigation of Space by Reactive Devices," in *Work on Rocketry by K. E. Tsiolkovsky,* NASA TT F-243, ed. M. K. Tikhonravov, G. I. Sedlenek, and T. N. Trofimova (Washington, D.C.: NASA, n.d.), 217.
11. Robert H. Goddard, "The Ultimate in Jet Propulsion" and "The Great Migration," *Papers of Robert H. Goddard* 3:1611.
12. Ibid. 3:1612.

Selected Bibliography

No one stands in greater debt to the scholars who preceded him than the author of a synthetic book. Although the volume you hold in your hands rests on considerable research in original materials conducted over a period of two decades, it is principally based on what those who came before me have discovered and published.

The chapter notes provide the source of all quotations. This bibliography lists books that helped to shape the text. Because many of the books were of value to more than one chapter, they have been grouped in larger chronological units.

General

Anderson, Frank W. *Orders of Magnitude: A History of NACA and NASA, 1915–1980.* Washington, D.C.: NASA, 1981.

    A solid, concise history of the agencies responsible for aerospace research and development since World War I.

Bilstein, Roger. *Flight in America, 1900–1980.* Baltimore: Johns Hopkins University Press, 1984.

    The history of America in space within the context of the age of flight.

Clark, Arthur C., ed. *The Coming of the Space Age: Famous Accounts of Man's Probing of the Universe.* New York: Meredith Press, 1967.

    First-person accounts of the early history of the space age by those who were present.

Emme, Eugene. *A History of Space Flight.* New York: Holt, Rinehart and Winston, 1965.

    A concise history of the early space age by the first head of the NASA history program.

Launius, Roger. *NASA: A History of the U.S. Civil Space Program.* Melbourne, Fla.: Krieger Publishing, 1994.

Ley, Willy. *Rockets, Missiles and Men in Space.* New York: Viking Press, 1968.

    A genuine pioneer of the early space age, and its finest chronicler, Willy Ley published a thin volume entitled *Rockets* in 1944. The book grew longer and more detailed as it passed through four editions over the next quarter of a century. The final edition remains the most readable introduction to the history of rocket technology and the prehistory of space flight.

Logsdon, John, ed. *Exploring the Unknown: Selected Documents in the History of the U.S. Space Program.* Washington, D.C.: NASA, 1995.
    The ultimate book of readings on the history of space policy.
McDougall, Walter A. . . . *the Heavens and the Earth: A Political History of the Space Age.* New York: Basic Books, 1985.
    This Pulitzer Prize–winning volume traces the roots of the space age from the utopian dreams of a handful of late nineteenth- and early twentieth-century visionaries to the state-sponsored space race between two technocratic giants and the economic, political, and cultural changes that it produced.
Roycroft, Michael. *The Cambridge Encyclopedia of Space.* Cambridge: Cambridge University Press, 1989.
    A single volume offering a solid, well-illustrated coverage of the history of the subject and the state of the art up to the time of publication.
Thomas, Shirley. *Men of Space.* Philadelphia: Chilton, 1960.
    A multivolume series of biographical sketches.
von Braun, Wernher, and Frederick I. Ordway III. *History of Rocketry and Space Travel.* New York: Thomas Y. Crowell, 1975.
    The best known engineer/manager of the space age cooperated with his friend and fellow engineer, Frederick I. Ordway III, on a number of books, including this volume, which remains the standard illustrated history of the subject.
Winter, Frank. *Rockets into Space.* Cambridge: Harvard University Press, 1990.
    A concise, comprehensive, and fairly detailed history of the rocket from an authority in the field.
Special note should be taken of series publications: *NASA Historical Pocket Statistics,* published annually by NASA Headquarters, and the *TRW Space Log,* no longer published, offer an annual quick reference of dates, times, and place of international space events. The series of published papers offered at the meetings of the International Academy of Astronautics, although seldom analytical, include a broad range of papers, from autobiographical notes by astronautical pioneers to technically detailed papers by engineers, enthusiasts, and historians. The first volume in the series was published as F. C. Durant III and George S. James, eds., *First Steps toward Space.* Washington, D.C.: Smithsonian Institution Press, 1974. All subsequent volumes, edited by a variety of individuals, have been published under the title *History of Rocketry and Astronautics.*

The Prehistory of Space Flight

Aldiss, Brian W. *Billion Year Spree: The True History of Science Fiction.* New York: Schocken Books, 1974.
Baumgart, Carola. *Johannes Kepler: Life and Letters.* New York: Philosophical Library, 1951.
Ferris, Timothy. *Coming of Age in the Milky Way.* New York: Doubleday, 1988.
Guthke, Karl S. *The Last Frontier: Imagining Other Worlds, from the Copernican Revolution to Modern Science Fiction.* Ithaca, N.Y.: Cornell University Press, 1983.
Hall, Marie Boas. *Nature and Nature's Laws: Documents of the Scientific Revolution.* New York: Harper & Row, 1970.

Koestler, Arthur. *The Sleepwalkers: A History of Man's Changing Vision of the Universe.* New York: Grosset & Dunlap, 1959.

Koyre, Alexandre. *The Astronomical Revolution.* Ithaca, N.Y.: Cornell University Press, 1973.

Kuhn, Thomas S. *The Copernican Revolution.* Cambridge: Harvard University Press, 1957.

Nicholson, Marjorie Hope. *Voyages to the Moon.* New York: Macmillan, 1960.

Sheldon, Charles. *Review of the Soviet Space Program.* Washington, D.C.: GPO, 1965–.

von Braun, Wernher, and Frederick I. Ordway III. *The Rockets' Red Glare: An Illustrated History of Rocketry through the Ages.* Garden City, N.Y.: Doubleday/Anchor Press, 1976.

The Pioneers

Bainbridge, William. *The Spaceflight Revolution: A Sociological Study.* Malabar, Fla.: Robert Krieger, 1981.

Dornberger, Walter. *V-2.* New York: Viking Press, 1954.

Goddard, Esther C., and G. Edward Pendray. *The Papers of Robert H. Goddard.* New York: McGraw-Hill, 1970.

Hunt, Linda. *Secret Agenda: The United States Government, Nazi Scientists, and Project Paperclip.* New York: St. Martin's Press, 1991.

Irving, David. *The Mare's Nest: The German Secret Weapons Campaign and Allied Countermeasures.* Boston: Little, Brown, 1964.

Korneev, L. K., ed. *Tsander: Problems of Flight by Jet Propulsion.* NASA TT-147. Jerusalem: Israel Program for Scientific Translations, 1964.

Lasby, Clarence. *Project Paperclip.* New York: Atheneum, 1971.

Lehman, Herbert. *This High Man: The Life of Robert Goddard.* New York: Fararr, Strauss & Giroux, 1963.

Lindbergh, Charles A. *Autobiography of Values.* New York: Harcourt Brace Jovanovich, 1976.

Moskowitz, Sam. *The Immortal Storm: A History of Science Fiction Fandom.* Westport, Conn.: Hyperion Press, 1974.

Neufeld, Michael. *The Rocket and the Reich: Peenemünde and the Coming of the Ballistic Missile Era.* New York: Free Press, 1995.

Ordway, Frederick I., III, and Randy Leiberman. *Blueprint for Space: Science Fiction and Science Fact.* Washington, D.C.: Smithsonian Institution Press, 1992.

Ordway, Frederick I., III, and Mitchell R. Sharpe. *The Rocket Team: From the V-2 to the Saturn Moon Rocket.* New York: Thomas Y. Crowell, 1979.

Ordway, Frederick I., III, and Ernst Stuhlinger. *Wernher von Braun: Crusader for Space.* Malabar, Fla.: Robert Krieger, 1994.

Rynin, N. A. *Interplanetary Flight and Communications.* NASA TT F- 646. Washington, D.C: NASA, 1971.

von Kármán, Theodore, and Lee Edson. *The Wind and Beyond.* Boston: Little, Brown, 1967.

Winter, Frank. *Prelude to the Space Age: The Rocket Societies, 1924–1940.* Washington, D.C.: Smithsonian Institution Press, 1983.

Yur'yev, B. N., ed. *Collected Works of K. E. Tsiolkovskii.* NASA TT F-237. 3 vols. Washington, D.C.: NASA, 1965.

Missiles and Moons

Baker, David. *The Rocket: The History and Development of Rocket and Missile Technology.* New York: Crown, 1978.

Burrows, William E. *Deep Black: Space Espionage and National Security.* New York: Random House, 1986.

Chang, Iris. *Thread of the Silkworm.* New York: Basic Books, 1995.

Clark, Phillip. *The Soviet Manned Space Program.* New York: Crown Books, 1988.

DeVorkin, David. *Science with a Vengeance: How the Military Created the U.S. Space Sciences after World War II.* New York: Springer-Verlag, 1992.

Emme, Eugene. *The History of Rocket Technology: Essays on Research, Development, and Utility.* Detroit: Wayne State University Press, 1964.

Green, Constance, and Milton Lomax. *Vanguard: A History.* Washington, D.C.: Smithsonian Institution Press, 1971.

Harford, James. *Korolev: How One Man Masterminded the Soviet Drive to Beat America to the Moon.* New York: Wiley, 1997.

Hickam, Homer H., Jr. *Rocket Boys.* New York: Delacorte Press, 1998.

Huzel, Dieter. *Peenemünde to Canaveral.* Englewood Cliffs, N.Y.: Prentice-Hall, 1962.

Medaris, John B. *Countdown for Decision.* New York: G. P. Putnam's Sons, 1960.

Newell, Homer. *Beyond the Atmosphere: Early Years of Space Science.* Washington, D.C.: NASA, 1980.

Schwiebert, Ernst G. *A History of U.S. Air Force Ballistic Missiles.* New York: Frederick Praeger, 1964.

Slukhai, A. I. *Russian Rocketry: A Historical Survey.* Moscow: Academy of Science of the USSR, 1965.

Stoiko, Michael. *Soviet Rocketry: Past, Present and Future.* New York: Holt, Rinehart and Winston, 1970.

Zaloga, Steven. *Target America: The Soviet Union and the Strategic Arms Race.* Novato, Calif.: Presidio Press, 1993.

From Mercury to Apollo

Air & Space Smithsonian 4, no. 2 (June–July 1989). Apollo commemorative edition.

Aldrin, Edwin. *Return to Earth.* New York: Random House, 1973.

Aldrin, Edwin, and Malcom McConnell. *Men from Earth.* New York: Bantam Books, 1989.

Armstrong, Neil, Michael Collins, and Edwin Aldrin. *First on the Moon.* Boston: Little, Brown, 1970.

Bilstein, Roger. *Stages of Saturn: A Technical History of the Apollo/Saturn Launch Vehicles.* Washington, D.C.: NASA, 1980.

Borman, Frank. *Countdown.* New York: William Morrow, 1988.

Carpenter, M. Scott, et al. *We Seven: By the Astronauts Themselves.* New York: Simon and Schuster, 1962.

Chaikin, Andrew. *A Man on the Moon.* New York: Viking Press, 1994.

Collins, Michael. *Carrying the Fire: An Astronaut's Journeys.* New York: Ballantine Books, 1974.

———. *Liftoff: The Story of America's Adventure in Space.* New York: Grove Press, 1988.

Cortright, Edgar M. *Apollo Expeditions to the Moon.* Washington, D.C.: NASA, 1975.

Ezell, Edward C., and Linda Ezell. *The Partnership: A History of the Apollo-Soyuz Test Project.* Washington, D.C.: NASA, 1978.

Hacker, Barton C., and James Grimwood, *On the Shoulders of Titans: A History of Project Gemini.* Washington, D.C.: NASA, 1977.

Hallion, Richard, and Tom Crouch, eds. *Apollo XI: Ten Years Since Tranquility Base.* Washington, D.C.: Smithsonian Institution Press, 1979.

Logsdon, John. *The Decision to Go to the Moon: Project Apollo and the National Interest.* Cambridge, Mass.: MIT Press, 1970.

MacKinnon, Douglas, and Joseph Baldanza. *Footprints: The Twelve Men Who Walked on the Moon Reflect on Their Flights and the Future.* Washington, D.C.: Acropolis Books, 1989.

Rosholt, Robert L. *An Administrative History of NASA, 1958–1963.* Washington, D.C.: NASA, 1966.

Schirra, Walter. *Schirra's Space.* Boston: Quinlan Press, 1988.

Sloop, John L. *Liquid Hydrogen as a Propulsion Fuel, 1945–1959.* Washington, D.C.: NASA, 1978.

Swenson, James, Jr. *Chariots of Apollo: A History of Manned Lunar Spacecraft.* Washington, D.C.: NASA, 1979.

Swenson, James, Jr., James Grimwood, and Charles C. Alexander. *This New Ocean: A History of Project Mercury.* Washington, D.C.: NASA, 1966.

Trenta, Joseph J. *Prescription for Disaster: From the Glory of Apollo to the Betrayal of the Shuttle.* New York: Crown, 1987.

Vladimirov, Leonid. *The Russian Space Bluff.* New York: Dial Press, 1973.

Toward Distant Horizons

Clarke, Arthur C. *The Promise of Space.* New York: Harper & Row, 1968.

Compton, L. W. David, and Charles D. Benson. *Living and Working in Space: A History of Skylab.* Washington, D.C.: NASA, 1983.

Corliss, Edgar M. *Scientific Satellites.* Washington, D.C.: NASA, 1967.

———. *Space Probes and Planetary Exploration.* Washington, D.C.: NASA, 1965.

Ezell, Edward C., and Linda N. Ezell. *On Mars: Exploration of the Red Planet, 1958–1978.* Washington, D.C.: NASA, 1984.

Fallaci, Oriana. *If the Sun Ever Dies.* New York: Atheneum, 1967.

Hall, R. Cargill. *Lunar Impact: A History of Project Ranger.* Washington, D.C.: NASA, 1977.

Heppenheimer, T. A. *Toward Distant Stars: Space as a New Frontier.* New York: Fawcett Columbine, 1979.

Hill, Janice. *Weather from Above: America's Meteorological Satellites.* Washington, D.C.: Smithsonian Institution Press, 1991.

Mack, Pamela. *The Politics of Technological Change: A History of Landsat.* Cambridge, Mass.: MIT Press, 1990.

McCurdy, Howard. *Inside NASA: High Technology and Organizational Change in the U.S. Space Program.* Baltimore: Johns Hopkins University Press, 1993.

———. *Space and the American Imagination.* Washington, D.C.: Smithsonian Institution Press, 1997.

NASA Advisory Council. *Planetary Exploration through Year 2000: Scientific Rationale.* Washington, D.C.: NASA, 1988.

Newell, Homer E. *Beyond the Atmosphere: The Early Years of Space Science.* Washington, D.C.: NASA, 1980.

Oberg, James, and Alcestis Oberg. *Pioneering Space: Living on the Next Frontier.* New York: McGraw-Hill, 1986.

Ride, Sally K. *Leadership and American's Future in Space: Report to the Administrator.* Washington, D.C.: NASA, 1987.

Tatarewicz, Joseph. *Space Technology and Planetary Astronomy.* Bloomington: Indiana University Press, 1990.

U.S. National Commission on Space. *Pioneering the Space Frontier: Final Report of the National Commission on Space.* New York: Bantam Books, 1986.

Index